Circuitbuilding Do-It-Yourself For Dummies®

D0127337

Resistor Color Code

Color	Value Stripe	Multiplier Stripe	Tolerance Stripe
Black	0	$\times 1$ (10^0)	
Brown	1	$\times 10$ (10^1)	1%
Red	2	$\times 100$ (10^2)	2%
Orange	3	$\times 1000$ (10^3)	
Yellow	4	$\times 10,000$ (10^4)	
Green	5	$\times 100,000$ (10^5)	0.5%
Blue	6	$\times 1,000,000$ (10^6)	0.25%
Violet	7	$\times 10,000,000$ (10^7)	0.1%
Gray	8	$\times 100,000,000$ (10^8)	0.05%
White	9	$\times 1,000,000,000$ (10^9)	
Gold		$\times 0.1$	5%
Silver		$\times 0.01$	10%
No color			20%

For more information on resistor markings: www.radio-electronics.com/ info/data/resistor/resistor_col_code.php.

Metric System of Units

Prefix	Symbol	Multiplication Factor
tera	T	10^{12}
giga	G	10^{9}
mega	M	10^{6}
kilo	k	10^{3}
centi	c	10^{-2}
milli	m	10^{-3}
micro	μ	10^{-6}
nano	n	10^{-9}
pico	p	10^{-12}

Voltage Conversions

Sine or square wave	$V_{PEAK\text{-}TO\text{-}PEAK} = 2 \times V_{PEAK}$
Sine wave	$V_{RMS} = 0.707 \times V_{PEAK}$, $V_{PEAK} = 1.414 \times V_{RMS}$
Square wave	$V_{RMS} = V_{PEAK}$
Power to decibels	$dB = 10 \log_{10} (Power\ 1\ /\ Power\ 2)$
Voltage to decibels	$dB = 20 \log_{10} (Voltage\ 1\ /\ Voltage\ 2)$
Decibels to power	$Power\ 1 = Power\ 2 \times antilog_{10} (dB\ /\ 10)$
Decibels to voltage	$Voltage\ 1 = Voltage\ 2 \times antilog_{10} (dB\ /\ 20)$

The antilog or inverse log function is often labeled $\log^{-1}$ on calculators.

Dimension Conversions

25.4 mm/inch	0.0393 inch/mm
2.54 cm/inch	0.393 inch/cm
30.48 cm/foot	0.0328 foot/cm
0.305 meter/foot	3.28 foot/meter
0.914 meter/yard	1.094 yard/meter

Capacitor Value Markings

###L (Three numbers and a letter)	Numbers 1 and 2 are value digits.
	Number 3 is a multiplier: $0 = \times 1$, $1 = \times 10$, $2 = \times 100$, $3 = \times 1000$, $4 = \times 10,000$.
	Letter denotes tolerance: J = 5%, K = 10%, L = 20%
##p or ##n	Numbers 1 and 2 are value digits.
	p denotes pF, n denotes nF.

For more information on capacitor markings: www.radio-electronics.com/info/data/capacitor/ capacitor-markings.php.

Circuitbuilding Do-It-Yourself For Dummies®

Cheat Sheet

Drill Sizes Commonly Used in Electronics

Size Number	Diameter	Next Largest Fractional Size	Clears Screw Size	For Self-Tapping Screw Size
11	0.191″	13/64″		10
19	0.166″	11/64″	8	
21	0.159″	11/64″		10-32
25	0.149″	5/32″		10-24
28	0.140″	9/64″	6	
29	0.136″	9/64″		8-32
33	0.113″	1/8″	4	
36	0.106″	7/64″		6-32
43	0.089″	3/32″		4-40
44	0.086″	3/32″	2	
50	0.070″	5/64″		2-56

Complete drill bit table: http://en.wikipedia.org/wiki/Drill_and_tap_size_chart.

Where to Find Component Data Sheets

Source	Web Site
Datasheet Archive	www.datasheetarchive.com
NTE (manufacturer of cross-reference parts)	www.nteinc.com
Open Directory Project	www.dmoz.org/Science/Technology/Electronics/Reference/Application_Notes_and_Data_Sheets
Datasheet Café (directory of manufacturer datasheet sites)	www.datasheetcafe.com

Vendors for Electronics Hobbyists

Jameco	www.jameco.com	Electronic components and supplies
Digi-Key Electronics	www.digikey.com	Electronic components and supplies
Mouser Electronics	www.mouser.com	Electronic components and supplies
MCM Electronics	www.mcmelectronics.com	Repair and replacement products
All Electronics	http://allelectronics.com	Discount electronic and electro-mechanical parts
Marlin P. Jones & Associates	www.mpja.com	Discount electronic and electro-mechanical parts
Ocean State Electronics	www.oselectronics.com	Electronic components, tools, test instruments, kits
RadioShack	www.radioshack.com	Basic selection of parts, lots of adapters and connectors
Ramsey Electronics	www.ramseyelectronics.com	Basic to advanced kits
Tower Electronics	www.pl-259.com	Connectors and adaptors
Velleman, Inc.	www.velleman.com	Components, tools, and test equipment

For Dummies: Bestselling Book Series for Beginners

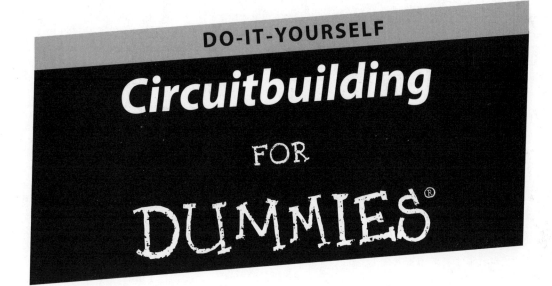

DO-IT-YOURSELF

Circuitbuilding

FOR

DUMMIES®

DO-IT-YOURSELF

Circuitbuilding

FOR

DUMMIES®

by H. Ward Silver

WILEY

Wiley Publishing, Inc.

Circuitbuilding Do-It-Yourself For Dummies®

Published by
Wiley Publishing, Inc.
111 River Street
Hoboken, NJ 07030-5774
www.wiley.com

Copyright © 2008 by Wiley Publishing, Inc., Indianapolis, Indiana

Published by Wiley Publishing, Inc., Indianapolis, Indiana

Published simultaneously in Canada

For general information on our other products and services, please contact our Customer Care Department within the U.S. at 800-762-2974, outside the U.S. at 317-572-3993, or fax 317-572-4002.

For technical support, please visit www.wiley.com/techsupport.

Wiley also publishes its books in a variety of electronic formats. Some content that appears in print may not be available in electronic books.

Library of Congress Control Number: 2007943806

ISBN: 978-0-470-17342-8

Manufactured in the United States of America

10 9 8 7 6 5 4 3 2 1

WILEY

About the Author

H. Ward Silver has the experience of a 20-year career as an electrical engineer developing instrumentation and medical electronics. He also spent 8 years in broadcasting, both programming and engineering. In 2000 he turned to teaching and writing as a second career. He is a contributing editor to the American Radio Relay League (ARRL) and author of the popular "Hands-On Radio" column in QST magazine every month. He is the author of the ARRL's Amateur Radio license study guides and numerous other articles. He developed the ARRL's online courses, "Antenna Design and Construction," "Analog Electronics," and "Digital Electronics." Along with his comedic alter-ego, Dr Beldar, Ward is a sought-after speaker and lecturer among "hams." When not in front of a computer screen, you will find Ward working on his mandolin technique and compositions.

Dedication

Circuitbuilding Do-It-Yourself For Dummies is dedicated to the many technical writers whose articles in *QST*, *Popular Electronics*, *73*, *CQ*, *Scientific American*, among others, inspired me to cut and solder and tinker my way through high school. Getting an amateur radio license on the way, that practical experience led directly to my first career as an electrical engineer. Another dedication is due my students and readers that make my second career as a writer equally enjoyable. If I can do for you what they did for me, I'll be very satisfied, indeed.

Author's Acknowledgments

In the early days of electrical experimentation, before "electronics" was even a word, there was no choice but to build one's own circuits. Back then, circuits were all about motors, lighting, and simple control systems. They were built with hammers, wrenches, screwdrivers, and, yes, soldering irons. Circuitbuilding was a full-body experience!

For a time not so long ago, it seemed that actually building one's own circuits was an activity that would go the way of AC-DC motor and knife switch. Electronic gadgets had become so inexpensive and easy to use, why should anyone bother to build anything more complicated than plugging cables together? The Internet and personal computer took building out of the physical world and into the realms of the network and cyberspace.

That trend has reversed in recent years. People of all ages are rediscovering the thrill and satisfaction of learning-by-doing. They've found that "lifting the hood" is just as much fun for electronics and circuits as developing a Web site or hooking up the latest gadget from the store. Not only just building, but modifying or "hacking" equipment, is providing hours of enjoyment, too!

If you're a budding circuitbuilder, welcome to the party! Join the thousands of ham radio operators, robotics enthusiasts, engineers, inventors, tinkerers, and hobbyists—people just like you. Heat up that soldering iron, turn on the voltmeter, and start building!

—H. Ward Silver

Publisher's Acknowledgments

We're proud of this book; please send us your comments through our online registration form located at www.dummies.com/register/.

Some of the people who helped bring this book to market include the following:

Acquisitions, Editorial, and Media Development

Senior Project Editor: Mark Enochs

Senior Acquisitions Editor: Katie Feltman

Senior Copy Editor: Barry Childs-Helton

Technical Editor: Kirk Kleinschmidt

Editorial Manager: Leah Cameron

Media Development Project Manager: Laura Atkinson

Editorial Assistant: Amanda Foxworth

Sr. Editorial Assistant: Cherie Case

Cartoons: Rich Tennant (www.the5thwave.com)

Composition Services

Project Coordinator: Lynsey Stanford

Layout and Graphics: Stephanie D. Jumper, Erin Zeltner

Proofreaders: Cindy Ballew, John Greenough

Indexer: Becky Hornyak

Publishing and Editorial for Technology Dummies

    Richard Swadley, Vice President and Executive Group Publisher

    Andy Cummings, Vice President and Publisher

    Mary Bednarek, Executive Acquisitions Director

    Mary C. Corder, Editorial Director

Publishing for Consumer Dummies

    Diane Graves Steele, Vice President and Publisher

    Joyce Pepple, Acquisitions Director

Composition Services

    Gerry Fahey, Vice President of Production Services

    Debbie Stailey, Director of Composition Services

Contents at a Glance

Table of Contents

Introduction

Perhaps you've never built anything electronic, and now you want to. Perhaps you have built something before, but now you want to do something different. Look no further. *Circuitbuilding Do-It-Yourself For Dummies* is the book for both kinds of readers. Primarily, this book is intended to act as an introduction and guide to someone just getting started with electronics and circuits. It covers basic tools and techniques. If you are somewhat experienced with electronics, you'll find the book most useful as a workshop reference for specific kinds of tasks. The latter half of the book focuses on specific how-tos: cables, connectors, measurements, and maintenance.

There are so many circuits and applications of electronics that it is impossible to provide a detailed how-to guide for even a tiny fraction of the different types! The goal of this book is to show you the tools and techniques that circuitbuilders use, common to a wide variety of electronic construction needs.

This book presents basic techniques most useful to beginners. As such, you won't find detailed discussions of advanced topics such as fabricating your own circuit boards or performing reflow soldering at home. Nevertheless, if you become familiar with the techniques in this book, it will be easier for you to move on to more sophisticated techniques. I'll also give you pointers about where to find information on them.

This book is *not* a circuit design course or cookbook. I'll be assuming that you already have a schematic from a book or magazine or maybe you've purchased a kit. This book shows you how to build it, not design it. The list of resources in Appendix A include quite a number of how-to-design books about electronics and even some online courses and tutorials.

What You're Not to Read

As you make your way through *Circuitbuilding Do-It-Yourself For Dummies,* feel free to skip around to where your interests and needs take you. You don't have to read each chapter in order. Use the Table of Contents or the Index to find help on a specific topic, such as wiring up a particular cable. The extensive Glossary in the back of the book will help with unfamiliar terms. Sidebars contain material that's interesting but not required reading.

Assumptions About You

The subject of electronics is big and broad and deep, but don't panic! You only need tackle the small steps at first — be comfortable and progress at your own speed. This book doesn't expect you to have an engineering degree or a complete shop. In fact, I deliberately performed all of the tasks myself with the simplest equipment and tools, just to be sure my readers could do them, too!

What I *do* assume about you, however, is that you're curious and motivated to build on the basic skills in *Circuitbuilding Do-It-Yourself For Dummies*. Take a few minutes to investigate the online resources I note throughout the book. You'll also find an extensive list of resources in Appendix A.

Finally, you don't have to run out and buy all of the tools and components shown in the book. I'm sure your local electronics emporium would love it if you did, but take your time! Each task lists the tools and materials needed, and you will be just fine if you acquire them as you need them.

How This Book Is Organized

Circuitbuilding Do-It-Yourself For Dummies is composed of six parts. You'll get started with some electronic construction basics, then move onto specific tasks that show how circuitbuilding is done. From there you can read about techniques that support circuitbuilding like taking measurements and maintenance. A Glossary and the famous Parts of Ten wrap up the book.

Part 1: Working Basics for Electronic-ers

This book doesn't neglect the basics — tools and techniques. You may have most of the tools, already! If you don't, this introductory part will help you get the ones you need. Then it's on to simple techniques for working with the materials you'll encounter when building circuits. I'll also help you read and make sense of electronic schematics, the language of circuitbuilders.

Part II: Building Circuits

This part of the book presents several ways of working with electronic components and materials to turn an idea into a living breathing circuit. By learning the basic techniques, you can build even the most complex circuits. Then learn how to install your circuit in a simple enclosure.

Part III: Cables and Connectors

Take a look at the back of any stack of electronic gadgets and what do you find? Cables and connectors! Lots of them! Yet the "how to" of making and repairing cables is rarely presented. That information doesn't get left out of this book! I cover all kinds of cables and connectors so that when your circuit is finally built, you'll be able to make the necessary connections to other equipment, too.

Part IV: Measuring and Testing

You can't see, smell, or touch electricity in your circuits — unless something goes pretty wrong! Testing and evaluating your circuits, then, takes some special electronic eyes and ears. This part of the book shows you how to use basic test instruments as part of the circuitbuilding process and during troubleshooting.

Part V: Maintaining Electronic Equipment

Circuitbuilding isn't just about soldering components together. Once you've built your circuit, what then? This part of the book covers installation and troubleshooting along with information on batteries and dealing with interference and noise. All of these topics are mighty handy out there in the Real World!

Part VI: The Part of Tens

Familiar to all *For Dummies* readers, these are condensed lists of helpful and (hopefully) memorable ideas. In this part, you'll get the top ten secrets of the art of circuitbuilding, as well as indispensable information on circuit first aid and some supplies you should keep handy.

Glossary

As you go through the book, specific technical terms in *italics* will often be found in the Glossary. Keep a bookmark in the glossary and you won't have to *gloss over* a term you don't understand.

Bonus Chapters

The book was so chock-full of critical info, we had to leave a few things out. But have no fear because you can find two bonus chapters on the Web site (`www.dummies.com/go/circuitbuildingdiyfd`) covering resistor and capacitor types.

Conventions and Icons

To make the reading experience as clear and uncluttered as possible, a consistent presentation style is used:

 Italics are used to note a new or important term.

 Web site URLs (addresses) use a monospace font.

Additionally you'll see the following icons used as markers for special types of information.

This icon alerts you to a hint that will help you understand a technical or operating topic. These are often referred to as "hints and kinks" by circuitbuilders.

This icon highlights a new term or concept that you'll need to know about. Be sure to check the book's Glossary, as well.

Whenever I could think of a common problem or "oops," you'll see this icon. Before you become experienced, it's easy to get hung up on some of these little things.

This icon lets you know that there are safety, rules, or performance issues associated with the topic of discussion. Watch for this icon to avoid common gotchas.

These icons remind you of an important idea or fact that you should keep in mind.

Where to Go from Here

If you are just getting started with electronics, I recommend that you read Parts I and II thoroughly and try a few of the tools and techniques. Building a kit (Chapter 4) is a great way to turn your newfound knowledge into a gadget you can really use — a great confidence builder! Then try a couple of the other techniques before striking out on your own. The tasks in Part III can be performed whenever they arise as you build circuits. Study the techniques in Parts IV and V and give them a try.

If you're more experienced with electronics and want to use this book as a reference and how-to guide, be sure to scan through the book first. I'll bet there are a few sections or tips that might be an "Ah, hah!" for you. The Table of Contents can serve as your reference for workbench use.

Appendix A lists many references and provides some bonus material about electronic components, too. Bookmark the sites you find most interesting or useful and you'll have an instant electronic reference library! The print references listed in Appendix A are those that I've found to have a long useful life — many can be found in used bookstores or online at a fraction of their new cost. Even older texts will provide excellent information about how circuits work.

I couldn't be more pleased to welcome all of you readers to the world of electronics and circuitbuilding. You'll be able to use these tools and techniques for a long time. Learning them launched me into a lifetime of professional electronics that I've found to be both rewarding and enjoyable. I hope it's the same for you!

Part I
Working Basics for Electronic-ers

In this part . . .

Are you ready to roll up your sleeves and get started? Well, the handiest place to begin is a tour of the toolbox and a review of a few techniques that every circuitbuilder must master. The better you are equipped and the more experience you have in building, the better you will be at this craft.

This part begins with a chapter that covers the physical tools that you'll need to create the circuits. Along with the hardware, you'll be introduced to some low-cost, easy-to-use software that makes circuitbuilding (and designing!) much easier.

And what book on electronics would be complete without a discussion of soldering? The second of these chapters introduces you to the fine art of melting solder. There's also some information about how to install your circuits in enclosures and on working with metal and plastics. Finally, get a handle on reading schematic diagrams — your roadmaps to understanding circuits everywhere!

Chapter 1

The Toolbox

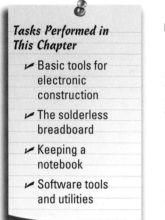

Tasks Performed in This Chapter

- ✔ Basic tools for electronic construction
- ✔ The solderless breadboard
- ✔ Keeping a notebook
- ✔ Software tools and utilities

To build anything, large or small, using the right tools makes a huge improvement in the quality of the finished product. The right tools will also speed up the process of building, minimize wasted materials, and reduce operator fatigue and stress. Sounds pretty important to have the right tools, doesn't it? You're right! This chapter shows you which, out of the zillions of tools, are the ones to use for building electronic circuits.

Basic Tools for Building Circuits

You'll be pleasantly surprised to find that you don't need a giant set of fancy tools to do excellent work! In fact, you may have most of them already and a couple of additional acquisitions are all that's needed.

Mechanically speaking, you'll need squeezers, cutters, turners, pokers, holders, and hole makers. That's pretty simple, isn't it? Of course, there is an incredible variety of available tools. I'll list the basic items you really need, ways to upgrade them, and some optional tools that are handy but not necessities. Then you go shopping!

Buy the best tools you can afford — always! Then take care of them — always! With care, tools will last a literal lifetime. The author's toolbox has perfectly functional and often-used tools that are 40 years old or more. Avoid bargain-bucket and no-name tools. An all-in-one tool is handy at times, but is no match for a single-purpose tool. Buy from a store with a no-questions-asked return policy that stands behind their tools.

The selection of tools listed in this section has been made with electronics in mind, not robot assembly, plumbing installation, or home wiring. Tools for those jobs are often inappropriate for the smaller scale of electronics. Conversely, electronic tools are often overmatched for beefier work. There is no one-size-fits-all tool selection!

The Klein Company has specialized in tools for electrical and electronic work for decades. They have an excellent selection of tools designed for every possible use at the electronics workbench. Their online catalog (`www.kleintools.com/Tool Catalog/index.html`) is a great reference. Klein is my favorite, but there are many other fine tool companies. Ace Hardware has a comprehensive introduction to many common types of tools on their Web site at `www.acehardware.com`. Click Projects⇨Solutions⇨Learning Guides to access the directory of informative pages.

Safety and visibility

Before you head off to the hardware store with a big list, be sure that right at the top you include some basic safety equipment — goggles (or safety glasses), workspace ventilation (for soldering smoke or solvent fumes), and first aid. Electronics may sound tame, but the first time you snip a wire and hear the sharp end "ping" off your safety glasses or take them off and find a small solder "splat" right in front of your eye, you'll be glad you had them on!

It sounds trite, but you really do need to be able to see what you're doing! There are two paths to seeing your electronics clearly; lighting and magnification. Your work-space simply has to be brightly lit, preferably from more than one angle to minimize shadows. Inexpensive swing-arm laps with floodlight bulbs are good choices because they can be moved to put light where you need it.

Head-mounted magnifiers are inexpensive and lightweight. The Carson MV-23 dual-power magnifier (`www.carsonoptical.com/Magnifiers/Hands%20Free`) is widely available and provides both x2 and x3 magnification. Swing-arm magnifiers, such as the Alvin ML100 (`www.alvinco.com`), can be positioned in front of your face and provide additional illumination, too. Magnifiers are often found at craft and sewing stores for considerably less cost than at office or technical-supply stores.

Pliers and tweezers

In the "squeezer" category are *pliers* and *tweezers*. The largest electronic thing you are likely to have to grab with pliers is a half-inch nut; the smallest will be tiny set screws. Pliers and tweezers that fit things in that range are good to have in your toolkit. Figure 1-1 shows a few examples of the pliers and tweezers that I use a lot.

The most common type of pliers are *slip-joint pliers* (8″) which have jaws that can be adjusted to grip large or small things. A small pair of *locking pliers* (6″) (optional) — also known as Vise-Grips™, come in very handy when working with connectors and can be used as an impromptu clamp or vise.

Needle-nose pliers (a generic term that covers many different styles of pliers) with serrated jaws are a necessity. You'll need a heavy pair of combination long-nose pliers (8″–9″, with or without a side cutter) for bending and holding. Smaller needle-nose pliers (5″–6″) will be used for positioning and holding delicate components. Additional pliers with extra-fine jaws (or bent-nose pliers) are nice to have in the toolbox, but not required.

Blunt-nose tweezers

Long-nose pliers   Fine-point tweezers   Slip-joint pliers

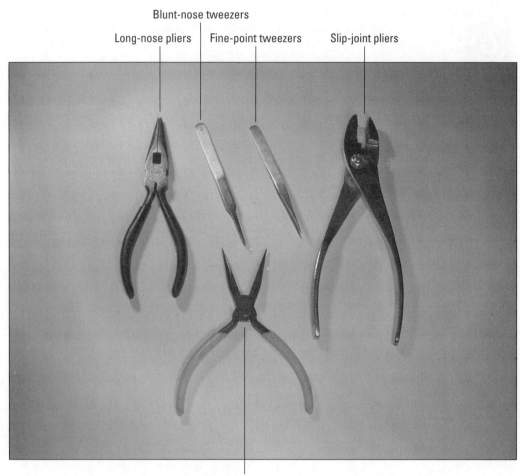

Needle-nose pliers

Figure 1-1: This set of pliers and tweezers will grab anything you're likely to encounter in electronics.

Tweezers are absolutely necessary when working with surface-mount devices (see Chapter 4) and small mechanical assemblies. They should be made of stainless steel; you'll need a pair with a blunt nose and a pair with pointed tips. Do not use regular bathroom or cosmetic tweezers — they're not really designed for electronics jobs.

Cutters and knives

Two pairs of wire cutters will suffice. For heavy wire, coaxial, and data cable, you'll need a pair of heavy-duty *diagonal cutters* (6″) like those in Figure 1-2. Get a pair with comfortable handles so that when you squeeze really hard you won't hurt your hand. For small wires, such as component leads, a 5″ pair of flush-cutting, pointed-nose or blunt-nose cutters is appropriate.

As you use your cutters day in and day out, they'll naturally lose their fine edge — although they may still cut wire just fine. For trimming very small wires, such as coaxial cable braid, you'll want a pair of very sharp cutters. It's a good idea to have one pair of "everyday" cutters and another pair used only for fine jobs — a miniature pair of pointed-nose cutters is good — and make sure those stay sharp.

A sharp knife is a must. For electronics-size jobs, a *utility knife* with a retractable segmented blade is a good choice. As the tip or edge dulls, you snap off the knife blade segment to expose new, sharp cutting edges.

Heavy scissors are used frequently and can even cut the lighter thicknesses of printed-circuit (PC) board. They will also be used to cut lighter gauges of sheet metal, such as aluminum and brass.

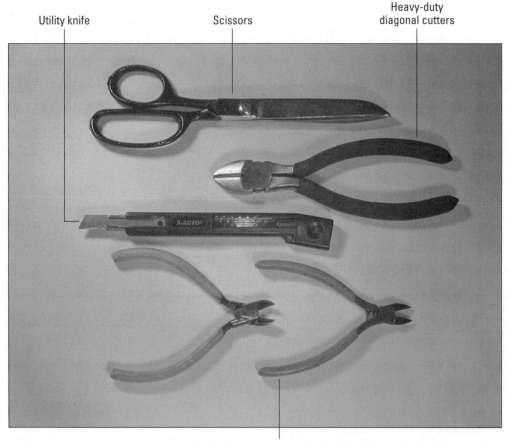

Utility knife   Scissors   Heavy-duty diagonal cutters

Miniature blunt-nose cutters and pointed-nosed cutters

Figure 1-2: The essential cutters and knives.

Screwdrivers and wrenches

Your toolbox should include both Phillips and flat-blade screwdrivers in sizes #0, #1, and #2. An optional long-shaft (8″ or longer) screwdriver is useful for getting at long cabinet-mounting screws in recessed locations. The many different types of screwdriver blades are explained and illustrated at `www.acehardware.com/sm-learn-about-screwdrivers--bg-1266832.html`.

A miniature flat-blade screwdriver with a 3/32″ blade will come in very handy as a general-purpose poker, pusher, and stirrer. It is particularly useful for mixing and applying epoxy! (Just don't let epoxy harden on the blade.)

Jeweler screwdrivers are handy, but not required. You'll use them mostly for attaching knobs to control shafts. If you do buy a set, make sure the shafts don't slip in their handles and that the blades are of good-quality steel. A lot of torque is applied to jeweler's screwdrivers; it's easy to twist off a blade or ruin an irreplaceable miniature screw if the blade isn't tough enough.

Obtain a set of *nutdrivers* for nuts from 1/4″ through 1/2″. These fit the nuts for screw sizes from #4 through 5/16″. The larger nutdrivers also fit switch- and control-mounting nuts. They will tighten the nuts without scratching a front panel and can be used on congested panels where a regular wrench can't be used.

Another optional tool is a miniature *Crescent® wrench* smaller than 6 inches long. Most mechanical fasteners used in electronics are too small for wrenches, but enough are large enough for the Crescent wrench to be a welcome sight in the toolbox.

A set of *Allen wrenches* is optional, but when you really need them (mostly for set screws), they have no substitutes. If you have a choice of buying a set of individual wrenches or a set mounted on a handle, the individual tools are somewhat easier to use (and lose). In addition, the ball-end wrenches can be used at an angle to the screw — which is sometimes necessary in tight quarters. Figure 1-3 shows several examples of screwdrivers, nutdrivers, and wrenches.

It is common for adjustable devices to come with an Allen wrench that fits their mounting set screws. When you're done installing the device, put the wrench in a locking plastic bag and label it with a permanent marker. You'll be able to find it much easier when the adjustment or mounting has to be redone later.

Drills and drill bits

To build electronic stuff, you'll need a small electric drill. A cordless model makes working on a car (or in the field) much easier, but cordless is not required. A 3/8″ chuck is big enough for electronic needs. A *hand drill* can be used on plastics, but is not recommended for general use. If you plan on installing your circuit in cabinets or project boxes with knobs or switches — especially with front panels that need to look good — invest in a small bench-mount *drill press*. It gives you dramatically improved ease of use and finished quality compared to what you get with a hand-held drill.

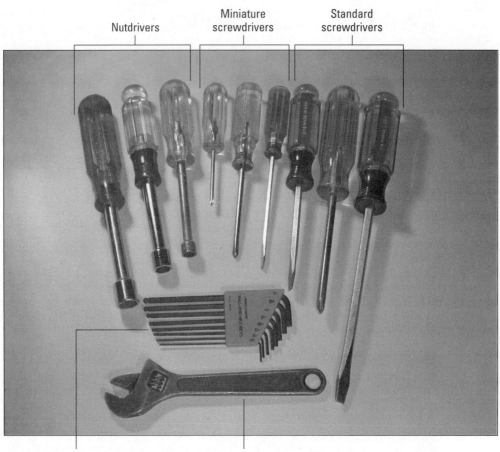

Figure 1-3: An assortment of screwdrivers is complemented by a set of nutdrivers. The miniature Crescent wrench and Allen wrenches round out the collection.

For delicate jobs, enlarging small holes, or just cleaning out a pre-drilled hole, a replacement drill chuck can make a fine hand-held holder for a drill bit. The machined metal chuck fits well in the hand and works like a handle for the bit; its size allows reasonable control of the bit.

You'll need an assortment of *drill bits* from 1/16″ to 3/8″. It's not necessary to have dozens of sizes and standard *twist bits* will suffice. A complete discussion of drill bit types and applications is available on the Ace Hardware Web site (`www.ace hardware.com`). Add an optional *countersink* bit to your collection of drilling tools to smooth the edges of holes.

While drilling small panels and enclosures, you should use a *vise*. For temporary and portable use, purchase a small *machinist's vise* or a small bench vise that clamps to the work surface. Trying to hold the material being drilled by hand often results in damage to your enclosure or panel — and if the material is seized by the drill bit, you can be injured. Examples of both can be viewed at `www.lexic.us/definition-of/machinist's_vise`.

It's important to mark a hole's center before drilling to prevent "walking" or wandering by the drill bit before the hole is deep enough to control the drill's position. A *center set punch* is tapped with a hammer, leaving a small dimple that can be placed precisely where the hole is to be drilled. Or you can use a nail, saving a bit of dough at the cost of a tiny bit of precision.

A *scratch awl* is handy for a number of punching and poking tasks. It can do the job of a center set in soft metal, plastic, and other soft materials. It makes holes in all sorts of flexible coverings. In wood, it can make a deep enough hole for a wood screw to be inserted.

A ½" *hand reamer* is used to enlarge a small hole. Using a reamer is often easier than drilling a large hole, especially in brittle plastics. An example showing how a reamer is used can be found in Chapter 5. *Needle files* come in a set including round, half-round, triangular, square, and other cross-sections (see Figure 1-4). They are used to smooth holes or file them into custom shapes.

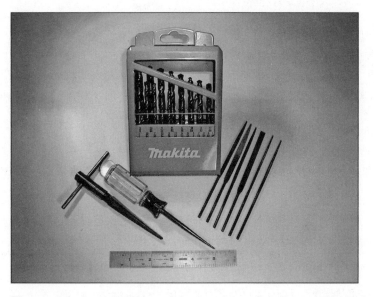

Figure 1-4: A set of drill bits and simple tools are all that is needed for basic electronic construction.

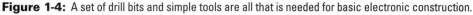

 The toolkits made by Kronus and Belkin include good, reasonable-quality starter tools. They are available from many electronics and tool retailers, including RadioShack, Sears, CompUSA, and others. You can replace individual tools with higher-quality selections as is convenient.

A somewhat odd, certainly optional, but very handy tool for circuitbuilders is the *nibbling tool*. All holes are not round! You may find that a display needs a rectangular cutout in a panel or that an elongated connector needs a rounded slot. Instead of drilling a lot of holes and then filing away (that works, but it takes a while), the nibbling tool shown at `http://adelnibbler.com/index.html` takes small bites out of sheet metal (and other thin material) in just about any shape you need!

Special electronic tools

As you put your circuitbuilding projects together, you'll find that you need a few specialized tools. You'll need some kind of wire stripper to remove insulation. A number of tools include wire-stripping capability, but they don't work as well (or as conveniently) as a tool made specifically for that purpose. The stripper should have individual positions for different sizes of wire, such as the Kronus 64-2980 available from RadioShack (www.radioshack.com). An automatic stripper (Kronus 64-2981) doesn't require pulling on the wire and is bulkier than the plier-like stripper — but it is fun to watch as it works!

Working on circuit boards and small devices is a lot easier if they are held firmly and at a convenient angle. The Panavise 301 vise shown in Figure 1-5 (www.panavise online.com/index.php) is made specifically for electronic and other detail work. The head of the vise swivels and turns 360 degrees. The *PC board vise head* has extra-wide jaws that can open wide for big boards.

Figure 1-5: The Panavise family of benchtop vises is designed for working with electronics and other small projects.

Some of the tasks later in this book require specific tools that do something unique — for example, the crimping tools used to install connectors (as shown in Part III of this book). Soldering equipment is covered in Chapter 2.

Measuring sticks

A small, metal *mechanic's rule* is a must-have in the electronics toolbox. Most are 6″ long with one side marked in metric units (mm and cm) and the other in English units (inches and fractions of inches). Because it's made of metal, it doubles as a conveniently firm straight-edge for marking or cutting. A short tape measure is also useful.

Stop giving me static!

As you peruse tool catalogs and Web sites, you'll see a number of accessories that dissipate static from people and tools. Why is this important? Well, if you've ever walked across a room and gotten a shock when you touched a doorknob, imagine that same amount of energy applied to a defenseless little transistor or IC! Suddenly, ESD (Electrical Static Discharge) protection starts to make sense!

A thorough introduction to ESD (www.esda.org/basics/part1.cfm) is published by the ESD Association, an electronics industry group that researches ESD protection. You can learn all about the different tools and techniques that prevent roasting your electronic components with a spark.

If you live in an area that is very dry on occasion, the best way to add ESD protection to your workspace is a static-dissipating mat and a personal grounding clip. Both of these connect to a safety ground and conduct excess charge away from sensitive electronics.

A permanent ruler is an option if your workspace allows. Use a yardstick to make permanent markings directly on the work surface. If you have a broken or cut tape measure, tack a length of the tape to the work surface. Being able to measure a cable or wire or other material without having to get out a new tool saves a lot of time!

Optionally, you may want to pick up a set of *calipers* to measure inside and outside widths and diameters, thicknesses, and even depths. Excellent quality calipers are available for a few dollars if you learn to read a vernier scale as instructed at www.marylandmetrics.com/tech/calipuse.pdf.

The Solderless Breadboard

One of the keys to learning about electronics is convenience. That is, learning and experimenting and testing should be as easy as possible. One way to make it easy is to use tools and techniques that reduce expense and bother. An excellent example of such a tool is the *solderless breadboard*. Using a breadboard is one of the basic starting points for the design of many types of circuits and projects. Also known as a *plugboard* or *prototyping board*, this miniature workbench allows you to whip up a circuit or try a new design in just minutes!

Using a breadboard

Figure 1-6 shows two examples of breadboards available from electronics parts and tool vendors. You can probably pick one up at your local RadioShack store. Models are available from postage stamp-sizes used for trying small circuits inside equipment all the way to foot-square models on which entire complex circuits can be built. A small one will do just fine as you start out, but it's a good idea to buy one size bigger than you think you need. You'll find yourself quickly outgrowing it, otherwise.

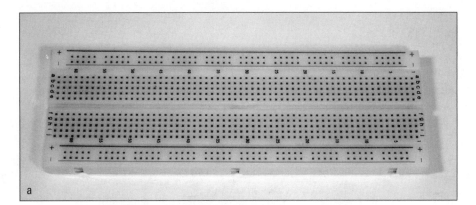

Figure 1-6: Two examples of solderless breadboards available from electronic retailers.

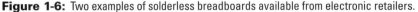

A solderless breadboard consists of plastic strips with small holes into which the leads of electronic components are inserted. (Figure 1-7 is a simplified drawing of a breadboard.) Brass strips under the holes connect each short row of openings together. Any two leads inserted into the same row of holes will be connected together electrically. The plastic body keeps adjacent strips from shorting together.

Up to four leads can be connected together in this way. If more common connections are required, a short piece of wire can be used to connect two (or more) rows together, creating a common electrical contact between all the holes in those rows. The slot between halves of the plastic strip is an insulating gap between the two sides so that integrated circuits with a DIP (Dual In-line Package) can be inserted with one row of pins on each side of the strip.

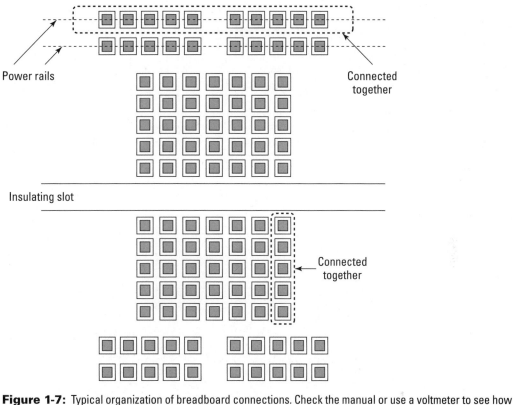

Power rails

Connected together

Insulating slot

Connected together

Figure 1-7: Typical organization of breadboard connections. Check the manual or use a voltmeter to see how your contacts are organized.

Most breadboards have areas for point-to-point circuit wiring and areas for distributing power and ground. These are called *rails* and run the length of the breadboard's plastic strips. For analog circuits, these are generally used for positive and negative power supplies, plus a common ground or return to the power supply. Builders of digital circuits that operate from a single voltage find it easier to "double up" and use the extra rail for a duplicate power-supply connection. Breadboards with more than one strip, each with its own set of rails, are easy to use for circuits that have both analog and digital circuitry.

If you are just getting started, you might consider purchasing a breadboard that comes with its own power supplies and possibly even some limited test capabilities, such as the Jameco 1537264 (www.jameco.com). More expensive models even have test meters and test signal generators.

While separate power supplies and test equipment might be more flexible and have additional features, the convenience of always having the test equipment connected and ready (remember?) will be appreciated.

Figure 1-8 shows some typical components inserted into the breadboard, ready to be "wired up." While circuits can be easier to build and troubleshoot with all the components laid horizontally, this generally isn't required. Here short pieces of solid wire make the connections from point to point around the circuit. Don't use stranded wire; the strands will move apart and cause hard-to-find short circuits.

What is a breadboard anyway?

Back in the old days, breadboards were literally just that — a wooden board on which loaves of bread were cut. Early electronics experimenters knew that these breadboards didn't conduct electricity (much), wouldn't catch fire (usually), and were cheap (definitely). That made breadboards just the right base for building a circuit — which in that era meant vacuum tubes: relatively high voltages and rather large components.

Many a wireless set or amplifier was constructed with sockets and terminal strips screwed to the soft wood of the kitchen breadboard! Although it's unlikely that you'll be slicing any loaves on modern breadboards, the name has stuck. In fact, the term *breadboarding* has come to mean the "roughing out" or "prototype" stage of designing and building electronic devices.

Power rails   Components

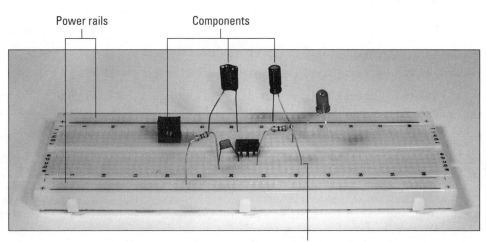

Lead inserted in hole

Figure 1-8: Component leads are inserted into the breadboard holes. Strips of contacts under the holes allow other components to be connected at the same point.

Breadboard materials

In keeping with the theme of convenience, breadboards hardly need any special materials to use! You'll need some test equipment to power and measure your circuits, certainly, but aside from the components themselves, little is needed. Here is a list of things you'll need:

✔ **Insulated jumpers (20- to 24-gauge solid, insulated wire in various colors):** It doesn't have to be *tinned* (coated with solder); bare copper is fine. A good source of suitable wire is scrap lengths of 4-conductor telephone wiring cable using for wiring the wall jacks (*not* the flat cable used to connect phones and wall sockets).

✔ **Bare jumpers (20- to 24-gauge solid bare wire):** This is used to connect adjacent rows of contacts, to create connection points for external equipment, or make leads for items that don't have suitable leads for insertion into the breadboard sockets. Save the clipped-off pieces of component leads to create a bountiful supply!

✔ **Leaded components:** It's very difficult, if not impossible, to use surface-mount technology (SMT) components with a breadboard. Make the task easier by purchasing and stocking only *leaded* components.

That's it! No special tools other than needle-nose pliers and a small pair of wire cutters are needed. You may also want to augment your eyesight by purchasing a pair of head-mounted magnifier glasses from a local craft store for a few dollars.

Limitations of breadboards

The breadboard sounds like a perfect way to build circuits, doesn't it? There are limits, however — and you should keep them in mind.

Current and voltage limits

The small contacts in a breadboard mean that they can only handle so much current before they are damaged by heating. Check the manufacturer's specification on how much current is safe. Higher currents can also melt the plastic strips. High voltage is often a problem, too, since the plastic insulation is only so thick. Arcing can also damage a breadboard. Whether from excessive voltage or current, damaged breadboard contacts can't be used reliably — and can't be repaired. A good rule of thumb is to limit breadboards to circuits that use a maximum of 100 mA and 50 V. If your circuit uses higher currents and voltages, it's a good idea to change your building methods or construct a separate circuit that only makes low-current connections to the breadboard circuit.

Frequency limit

The convenience of having lots of contacts and connections made of small wires has a drawback in poor performance for high-frequency signals. At high frequencies, the wires start to look like small inductors, upsetting circuit performance. Further, the many rows of closely spaced contacts act like small capacitors. Both the inductors and capacitors affect circuit performance in unpredictable ways. It's also harder to create a good, solid ground connection for a circuit of any complexity that's built on a breadboard. Another good general rule is to limit your circuit's highest frequencies to about 500 kHz. For digital circuits, the clock-speed limit is 1 MHz. Above those frequencies, your circuit won't be behaving the same way it will in a final version built with better techniques.

Contact wear-out

If you are a frequent builder, you'll probably start wearing out the breadboard's contacts. For example, some of the contacts will loosen, weakening their grip on a lead or wire. This is hard to detect — and can lead to intermittent problems that are difficult to assess and fix. If a contact has been overheated or has an oversized wire stuffed into it, its grip on smaller wires is relaxed. The connection points at one end of a power rail are particularly prone to this problem. Since you can't repair those contacts, it's best to mark which ones are bad and not use them again.

Your Notebook

The most important tool isn't one that lives in your toolbox, it's the one between your ears! The sharpening and lubricating for this tool comes from a notebook. Almost any old notebook will do — even one with cartoon characters on the cover. While a notebook filled with graph paper is the best, regular old lined or blank paper is fine. The important thing is to have a handy place to write down information as you work on projects.

Your notebook can be a record for design ideas, construction and installation notes, test results, project ideas — anything that you think goes in the notebook *should* go in the notebook. Believe me, you'll be a believer when you can go back into a years-old notebook and quickly find just the right circuit or look up the color code of a control cable you installed way back when!

Make a habit of opening the notebook before you even start work!

Software Tools

Can software be a tool for building electronics? Sure it can! If you can draw it on paper or calculate it, there is a software tool to help with the job. The only thing software can't do (yet) is fire up the iron and melt solder on that PC board. That's still your job, but by using the appropriate software, what you build will be finished faster and work more like what you expected.

There are far too many programs to try or even list, so only a few are mentioned here. More software is available all the time. If you do an Internet search for *"free electronic design software"* you'll be directed to Web sites such as the University of Nebraska's Electrical Engineering Shop page (eeshop.unl.edu/cad.html) or Technology Systems (www.tech-systems-labs.com/freesoftware.htm). They list many, many programs for you to try. Experiment and choose the ones you like!

Schematic and PC board layout

The actual term for the software with which you draw schematics is *schematic capture*. Software you can use to lay out your own circuit boards is *PCB layout*. The following packages listed here include both functions. While professional packages can cost thousands of dollars, there are some capable packages available for free or at very low cost. Free versions are usually limited in how many *pins* (meaning IC pins) can be used — and the designs may not be used for commercial purposes. For a beginner in circuitbuilding, these versions are just fine! Here are a few:

- **Easy PC** (www.numberone.com)
- **Dip Trace** (www.diptrace.com)
- **Eagle** (www.cadsoft.de; click Freeware)
- **Designworks Lite** (www.capilano.com/dwlite.html)

There are also low-cost PC board fabricators that provide schematic-capture and layout software (Express PCB, `www.expresspcb.com`) but they are usually proprietary packages that don't let you interface to other fabrication services. Nevertheless, this might not be a problem if such a package suits your purposes.

If you are familiar with PowerPoint software and only want to draw schematics that look good without any advanced features, a free package of schematic symbols developed by the author is available from the American Radio Relay League's Technical Information Service at `www.arrl.org/tis/info/HTML/Hands-On-Radio`.

Electronic simulators

The power of the PC is really put to work in electronic circuit *simulators* that can predict how your circuit will work. With a simulator, it is possible to do almost all your developmental work at the computer — and only turn on the soldering iron for the final version. To be sure, there are many subtle factors in circuit design that a computer doesn't know about or can't handle well, but these are well beyond what a beginning circuitbuilder worries about.

Simulators are powerful programs; they have a steep learning curve when you get beyond simple simulations. Nevertheless, there's no time like the present to try them out! These two packages are evaluation versions of professional-level circuit simulators:

✔ **Micro Cap** (`www.spectrum-soft.com/index.shtm`)

✔ **Intusoft ICAP** (`www.intusoft.com`)

The Linear Technology software, LTSPICE, is a capable version of the public-domain circuit-simulator program, SPICE. It's completely free from `www.linear.com/designtools/software/index.jsp` and also includes a switching power-supply design package.

Mechanical drawing software

It's also important to be able to make accurate drawings of panel layouts and other mechanical parts that are part of your project. Software that does mechanical drawings is called *CAD* for Computer-Aided Drafting. There are many inexpensive or free software packages (enter "cad drawing freeware" into an Internet search engine) for the downloading. Here are some general purpose drawing packages to try:

✔ **Vector Engineer** (`www.vectorengineer.com`)

✔ **CadSted** (`www.cadstd.com`)

There are also software packages for specialized drawing applications:

✔ **Scale** (`http://stiftsbogtrykkeriet.dk/~mcs/Scale.html`) is a Web application to design meter scales and control dials. You enter the data for your scale and it sends you a graphic file you can edit or print.

- ✔ **Dial and Panel** (http://hfradio.org/wb8rcr) are simple programs to make dial scales and design front panels.

- ✔ **Gpaper** (http://pharm.kuleuven.be/pharbio/gpaper.htm) draws any kind of graph paper you can think of!

Utilities and calculators

Literally thousands of utility software packages are available on the Internet. If you need one for a specific purpose, just type the purpose plus "design utility" into an Internet search engine — for example, "555 timer design utility" or "555 timer design calculator" — and dozens of programs and Web sites pop up. Caveat emptor (or, in this case, browser), of course; you don't know the pedigree of these programs. There is also a nice listing of electronic calculator programs at 101science.com/Radio.htm#Calculators.

As you collect the URLs for online calculators, set up a folder in your browser's Favorites list specifically for calculators. That way they'll always be easy to find.

You don't have to download every calculator individually as there are some very nice packaged sets. Here are two of my favorites, both free:

- ✔ **Hamcalc** (www.cq-amateur-radio.com/HamCalcem.html) has dozens of routines for all sorts of electronic design tasks.

- ✔ **Convert** (http://joshmadison.com/software/convert) is a terrific little utility that I leave on my PC desktop for whenever I have to convert a value between units of measure — say, barrels to pecks. Seriously, this is one of those tools that occasionally saves a whole lot of time.

Chapter 2

Basic Techniques

Topics and tasks in this chapter

- ✔ Metalworking
- ✔ Making a practice panel
- ✔ Learning to solder
- ✔ Desoldering
- ✔ Reading and drawing schematics

The mechanical part of electronic construction is usually quite straightforward. Circuits and their operating controls and connectors need to be mounted in a protective enclosure or housing and the circuits themselves need to be constructed properly. In this chapter, you'll be introduced to the basics of working on enclosures and panels made of sheet metal and plastic.

Soldering is a fundamental skill to the electronic technician. If you've never soldered, now is the time to learn. It's not hard and this chapter will get you started.

Finally, it's hard to work with electronics if you don't know the lingo of schematic diagrams. This chapter introduces the basics of schematics, both reading them and drawing them for others to read.

Basic Metalworking

Metalworking — it sounds so . . . so . . . industrial! Can't you just see the red-hot metal and sparks and anvils? Well, you don't have to heft a 16-pound blacksmith's hammer to do the kind of metalworking required for electronics. Most of the metal you'll encounter is light sheet metal, easily worked with hand tools no more complicated than a drill. You are actually more likely to work with plastic than metals. The same techniques apply to both materials.

Nearly all of the mechanical building for electronics that you'll do as a beginner will involve mounting components on an enclosure or a panel for an enclosure. For example, connectors, controls, indicators, and switches all need mounting holes. Good-looking panels can be made with common hand tools and patient attention to detail.

To show you how easy making a panel can be, the following task shows you how to make a practice panel. The material that you use can be just about any scrap piece of sheet; aluminum, or plastic, even PC board material. Metal should be no thicker than 16-gauge and plastic no thicker than ⅛″.

Making a Practice Panel

Stuff You Need to Know

Toolbox:
- Drawing software, center-set or scratch awl or nail, electric drill, ⅛" drill bit, ½" reamer, round needle file

Materials:
- Panel-mount variable resistor with ⅜" bushing (including mounting nut and lock washer)
- Miniature toggle switch (SW) with ¼" bushing
- T 1-¾ (5mm) LED
- Scrap or metal or plastic sheet, 2" × 4"

Time Needed:
Less than half a day

The key to getting your finished product to look good is to start with a carefully measured and marked layout. Rather than work directly on the panel, it's a lot easier to use software to lay out the controls and use a printed paper picture as a template. You can then mark the location of each hole or cut on the panel, using the template as a guide. The small amount of extra work at the beginning leads to higher-quality results and fewer mistakes.

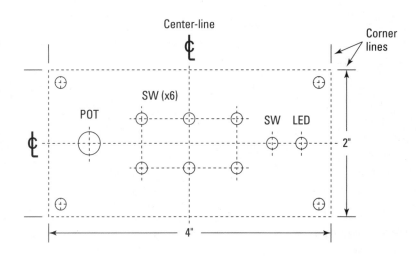

1. Use a drawing program (perhaps the program named "Panel" referenced in Chapter 1 or even PowerPoint). If you don't have drawing software, use graph paper and do an accurate job by hand. Place the corner mounting holes symmetrically at each corner. Locate and place the three holes on the horizontal center-line (the C with the L drawn through it is the symbol that denotes center-line). Locate and place the array of six ¼" holes (marked "¼ × 6") symmetrically around the vertical center-line. (Absolutely exact spacing isn't important on this practice panel.)

2. Print out a full-scale layout that is the actual size (4" × 2") of the panel. Place the finished layout upside down on a flat surface.

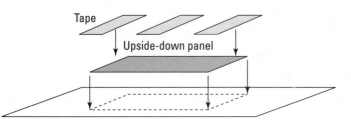

Upside-down layout

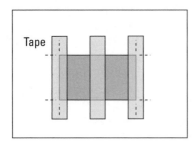

3. With the surface of the panel that you want to be the front facing down, place the panel on the back of the printed layout. Align the panel to fit between the corner marks. Use two or three strips of tape to hold the panel to the back of the layout.

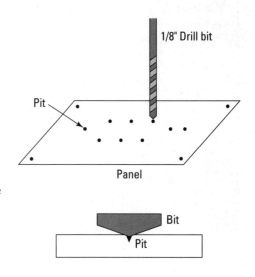

4. Turn the panel and layout over. Use the set, awl, or nail to make a small pit precisely where the center of each hole is marked on the paper; make the pit right through the paper. If you are using plastic, aluminum, or some other softer material, use only hand pressure to make the pit — don't use a hammer.

5. Remove the panel from the paper layout. You should clearly see a small pit on the surface of the panel.

6. Drill a ⅛″ hole at each pit, being careful to place the point of the drill bit directly in the pit. If you have a variable-speed drill, begin drilling slowly so that the drill bit starts the hole in the pit. The mounting holes in the corner of the panel will remain ⅛″. For larger holes, the ⅛″ holes are pilot holes to guide a reamer or larger drill bit.

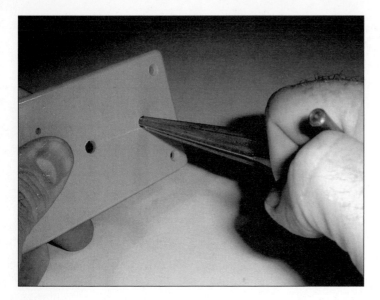

7. Use the reamer to enlarge each hole to fit the indicated component (POT, SW, or LED). Go slow until the indicated component (POT, SW, or LED) just fits through the hole. If your reamer won't fit into the ⅛" hole, use a ⁵⁄₃₂" or ³⁄₁₆" drill bit to enlarge the hole.

8. Deburr the holes by using a file (for the large hole) or a large drill bit (hold the bit in your fingers against the hole and turn it to take off any burrs or spurs).

9. Attach the potentiometer and switch, using a lock washer behind the panel and a finishing washer and nut on the front of the panel. You're done!

The Joy of Soldering

You can only use solderless breadboards (and wire-wrapping, as shown in Chapter 6) for so long! Soldering is the best way for the home circuitbuilder to make high-quality electrical connections between components and conductors. Learning how to solder is straightforward — all it takes is a reasonably steady hand and patience.

Soldering tools and materials

To do electronic soldering, you need a soldering iron and some solder, nothing more. The soldering iron should be a 30- to 50-watt, temperature-controlled model, as seen in Figure 2-1. This size iron is more accurately called a *soldering pencil* as the term "iron" includes everything up to giant 500-watt behemoths that could solder an I-beam. (Those are used in plumbing and sheet-metal work.)

Soldering stations

Soldering *stations* for electronics include the iron, a power supply, a holder for the iron, and a tip-cleaning sponge holder. Two good examples of soldering stations are the Weller WLC-100 (see Figure 2-1) and Hakko 936ESD. These stations and similar equipment are available from any of the vendors listed in Appendix A. Temperature is controlled manually with an adjustment on the station. The Weller WTPCT station uses an iron with special magnetic tips to control the temperature. The iron should have a tip temperature of 700° F.

Iron   Holder for the iron   Temperature control

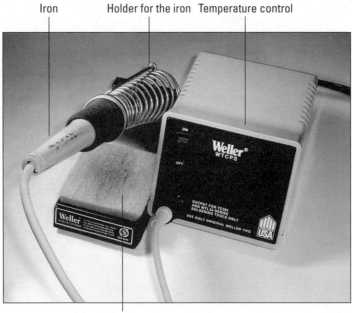

Tip-cleaning sponge

Figure 2-1: A soldering station: iron, temperature control, holder for the iron, and tip-cleaning sponge.

Is soldering toxic?

Solder contains lead (see the next paragraph on lead-free solder) and lead is toxic to humans. You would have to do a whole lot of soldering, however, to incur any significant hazard. The primary concern regarding solder is for people who solder every working day at their jobs — and for the amount of lead in electronic waste that could get into the water supply if disposed of carelessly. To be on the safe side, work in a well-ventilated area and wash your hands after you're done. Resist the temptation to hold the solder in your teeth when trying to make a complicated connection — it sounds crazy, but when you're trying to hold things together and solder at the same time, it might seem like a good idea.

The electronic industry is making the transition to lead-free solder, but hobbyists are not required to use it. There are several different types of lead-free solder, and the techniques for using it are more difficult than for lead-based soldering. The author recommends that you use lead-based solder until you are skilled enough to adopt the more advanced lead-free system.

What about soldering guns? Don't use them for electronics! Even a 100-watt soldering gun applies too much heat. These are intended for heavy wiring jobs and large connectors. If you try to use a soldering gun on a PC board, you'll quickly overheat and ruin the board's thin copper cladding along with the components you're trying to solder.

Soldering iron tips

Be sure you can use different tips on the iron you buy because there are several types of tips that you'll find useful. Most irons come with a ¹⁄₁₆" or ⅛" screwdriver tip. For regular electronics, you should have the following:

✔ A fine-point conical tip for small components and surface-mount soldering

✔ A ¹⁄₁₆" screwdriver tip for regular jobs

✔ A ⅛" screwdriver tip for heavier soldering

Solder and flux

The solder you need to use for electronics work is rosin-core flux, 60/40 or 63/37 alloy. (The two numbers in the fraction refer to the percentages of lead and tin, respectively.) Any thickness from 0.025" to 0.035" is fine for general use. You may want to have some thicker 0.050" for soldering large connectors and some 0.020" for delicate connections and surface-mount components. Purchase your solder from electronic suppliers; the type that's available from hardware and plumbing stores is too heavy for electronics.

Flux: A chemical that cleans metal surfaces to be joined by soldering or brazing. Rosin-core solder wire is actually like a tube made of solder that contains the rosin-type flux. Flux melts at a lower temperature than the solder, so it runs out of the solder strand and onto the metal surfaces before the solder melts. This prepares the surfaces and helps make a secure joint. The flux then boils away as smoke — that's what you smell when you're soldering. Flux can also be purchased separately as a liquid or paste for special soldering jobs.

Do *not* use acid flux; it's intended for use in plumbing. Acid flux is corrosive and will ruin any electronic project for which it is used!

Introduction to Soldering

Stuff You Need to Know

Toolbox:

✔ Soldering iron, solder

Materials:

✔ PC board and any component

Time Needed:

Less than an hour

The following task shows the basics of soldering a component lead to a circuit board. This is the most common soldering operation for most circuitbuilders. Once you're skilled at this operation, it's easy to solder other types of connections.

Before beginning, double-check: You DO have a pair of safety glasses or goggles, don't you? Solder can splash or splatter surprising distances. Protect your eyes! If you burn yourself with the soldering iron, the best thing to do is immediately run cold water over the burned area. A few minutes in cold water (or with an ice cube applied) will prevent most blisters. One more warning — DON'T grab for the soldering iron if you accidentally yank the cord or drop it. Chances are pretty good you'll grab something hot and that will result in a real burn. Just let the iron fall on the floor and pick it up. It's not hot enough to instantly start a fire, but it is hot enough to burn you!

1. Turn on the iron or soldering station and wet the sponge for cleaning tips. You don't need the sponge to be sopping wet, just a little beyond damp. An empty drinking-water bottle makes a good sponge wetter.

2. When the iron is hot (solder melts almost instantly when you touch it to the tip), wipe both sides of the tip across the wet sponge. Do this every time you take the iron out of the holder to keep it clean. If you are soldering many joints, clean the tip every few joints to prevent a residue of oxide and flux from building up.

Tinning: The coating of a metal surface with solder. The tin in the coating stays shiny and does not oxidize; solder will readily wet it. (Component leads are tin-plated for that reason.) Modern soldering irons have tips with a tough iron coating that doesn't need to be tinned the way the big irons for plumbing work have to be. It's enough to keep the tip clean; it will self-tin as the iron is used. After heavy use, the iron coating will eventually wear through, exposing the copper underneath. When this happens, it is best to just replace the tip; at that point, the copper will rapidly degrade and pit.

3. Make sure the component lead and pad are both free of grease, insulation, oxide, or any liquids. The metal surfaces should be bright and shiny.

4. Apply the tip to BOTH the lead and the pad. The easiest way to do this is to lay the slanting surface of the tip directly against the pad and press the side of the tip against the lead coming through the hole.

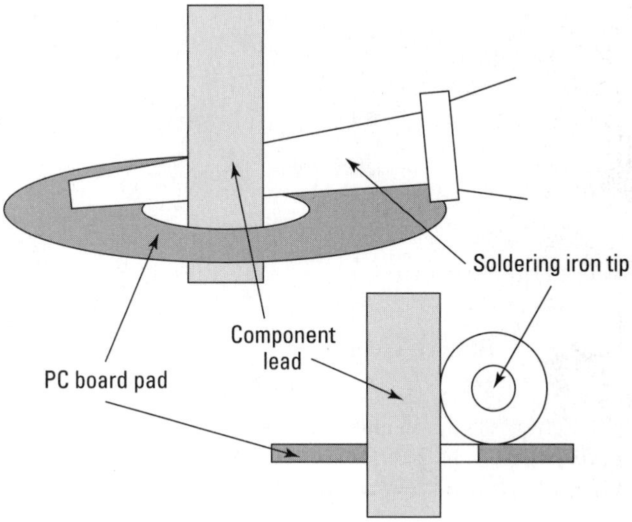

Soldering iron tip

Component lead

PC board pad

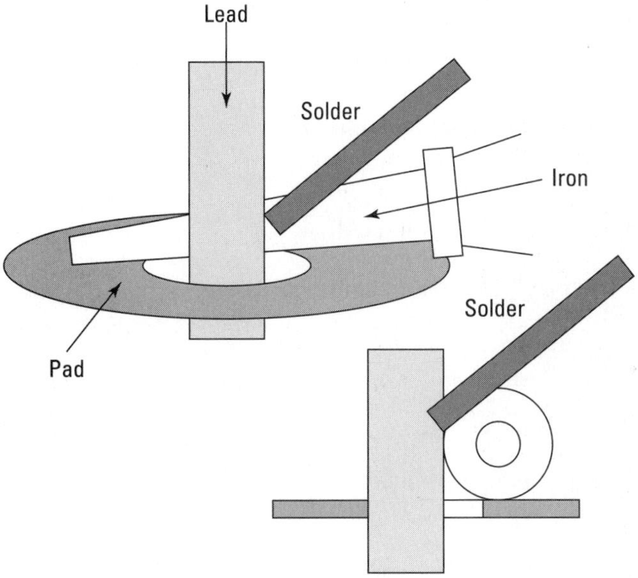

5. Apply the solder at the junction of iron, lead, and pad. It should melt almost immediately.

Lead

Solder

Iron

Solder

Pad

6. When the solder melts, feed a small amount onto the joint, then pull the solder strand away. You should use no more than ¼" of the strand on most joints.

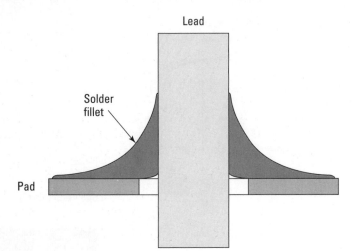

Lead

Solder
fillet

Pad

7. Leave the iron on the joint for no more than 1–2 seconds. The solder should flow around the lead, forming a *convex fillet* (solder fills the joint between the pad and lead) that completely encircles the lead and flows completely over the pad.

8. Remove the iron and let the joint cool. The fillet should be shiny. Inspect the pad to be sure excess solder hasn't flowed to adjacent leads or pads. You are looking for three types of defects:

- **Dull, rough, or pitted surface.** This indicates that the solder was over-heated or not heated enough. Sometimes adding a small additional amount of solder will allow the fillet to form properly — just re-melt the solder and add a small amount from the strand.

- **The joint is blobby or the solder pulls away from a lead or pad.** Either the lead or pad was not sufficiently heated or its surface was not clean. Remove the solder as described below and re-solder the connection.

- **Lead moves when pulled or bent.** This can also be an indication of the previous problem or the lead may have been moved before the solder solidified. Reheat the joint and try again — you may have to add a bit more solder.

Learning to solder with a kit

If you prefer detailed step-by-step instructions, there are a number of "learn-to-solder" soldering kits. Two of the more popular are these:

- ✔ **Elenco AK100** (www.pololu.com/products/elenco/0321). This kit even includes the soldering iron and a pair of wire cutters! (You'll want a better iron for long-term use, but this one is fine for beginners.)

- ✔ **Chaney Electronics C6880** (www.chaneyelectronics.com/products/learn-to-solder/c6880.htm). Four small PC board kits are provided in this package. You will have to provide your own soldering iron.

Desoldering

To "undo" a solder joint and remove the component lead takes a couple of steps. First, you remove excess solder with a *solder sucker*. This is a spring-loaded tool whose tip is placed against the solder joint. You melt the old solder with the soldering iron, put the tip of the solder sucker next to the bead of melted solder, and then release the spring in the tool: A piston retracts rapidly inside the solder sucker's tip, "inhaling" the molten solder into the tool. You may have to repeat this action a couple of times to get the bulk of the solder out. (Go to www.action-electronics.com/ desolder.htm for a look at these interesting tools.)

Since a solder sucker usually can't remove every bit of solder from a hole in a PC board, the next step is to use *desoldering braid*. The braid (also called *wick*) is made of fine copper wire impregnated with solder flux (that's what makes it sticky). When you press this wire onto a soldered connection with a soldering iron (as seen in Figure 2-2), the heat melts the flux, and then melts the solder; capillary action draws the molten solder up into the braid and away from the lead and the hole. Then you can use needle-nose pliers to wiggle the lead a bit, breaking whatever tiny bits of solder remain, after which you can remove the lead.

Figure 2-2: Use desoldering braid to remove solder from a joint so you can take a component lead out of a hole.

Making Sense of Schematic Diagrams

Once upon a time, schematics were hand-drawn by skilled drafters on large sheets of vellum drawing paper, reproduced on ammonia-fuming blue-line diazo paper, and laid out flat in special cabinets. Today, most schematics are entered into a computer, kept as files, and transferred at the speed of information around the Internet.

Just because a computer program was used to make a schematic doesn't mean that it's correct — or even that it makes sense! Computers are happy if all the lines connect and fit on a page, but the human reader needs more. Before beginning a project with a schematic from an unknown source, you should make sure that it makes sense. Check the Web site or later issues of the magazine to see if there are any corrections.

A schematic diagram has three purposes:

- ✔ Serves as a symbolic connection diagram (not a construction diagram)
- ✔ Describes the circuit's design
- ✔ Helps with troubleshooting

Think of the schematic as a roadmap to your circuit. By using standard symbols that are the equivalent of an electrical alphabet, schematics store complex design information in a form that can be read by others who know the symbols. The connections between the symbols, when arranged clearly, create small groups of symbols describing specific functions, such as an amplifier, a filter, or a control circuit. To an experienced reader of schematics, these groupings make it clear how the circuit's many parts work together. This also allows someone trying to improve or troubleshoot the circuit to understand how the circuit is *supposed* to work.

The history of schematics

As electrical systems became more and more complex around 1900, engineers needed more and more sophisticated methods of describing them. The first schematics were nothing more than pictorial diagrams showing the physical location of each component. These weren't sufficient for really complicated wiring, and were soon replaced by ladder diagrams that showed component symbols connected between parallel power *busses* or *rails*. Ladder diagrams are still used today, as you can see in the wiring diagrams of appliances like washers and dryers. The terms *bus* and *rail* are still used

to refer to shared power connections, even though they may not be the long, straight conductors they once were.

By 1930, nearly 50 years after the use of electricity became commonplace, schematics (the word was invented in 1894) and many of the symbols used today had become accepted standards. Standardized schematics were crucial to electrical industries because a visual description was the only way to share the complex information necessary to build complex equipment. Today, computers don't need schematics, but people still do.

Reading a Schematic

Before tackling a schematic, you should become familiar with the symbols and conventions used to draw schematics. A good set of symbols can be found online at www.kpsec.freeuk.com/symbol.htm. You can download the symbols as graphics files from this page, as well.

Let's say you're trying to decipher a new schematic, such as the period-to-voltage converter shown in the following figures. Try to break the schematic into sections; this will help the individual details make sense. It works for any kind of circuit, provided the schematic has been drawn in a reasonably organized way.

Don't be concerned that the circuit in this task is complicated. The circuit is presented only as an example of how to break up a busy schematic into bite-sized sections that you can analyze individually. A comprehensive treatment of reading and drawing schematics is beyond the scope of this book.

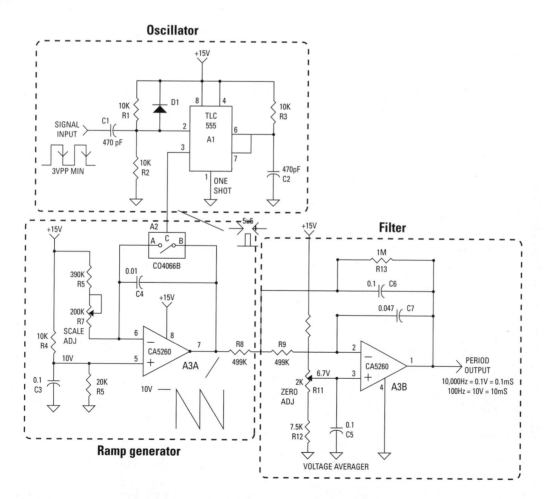

1. Make a few photocopies of the schematic. Writing notes, labeling parts and pieces, and outlining sections on the schematic, especially in different colors (red for power, green for signals, blue for control signals, and so on), helps keep track of all the functions.

2. Scan the entire schematic and try to identify the major functions and components. Draw lines around the parts of the circuit that perform related functions. If necessary, make your own block diagram. Browse the entire schematic, identifying the major functions and important components. Draw lines around the sections that have a common function, such as the power supply, an output amplifier, filters, sensors, and so on.

3. When you have a good working grasp of what the circuit does and how the schematic is organized, start focusing on the details. If you need to, make more copies of the schematic to keep everything straight. Remember that every schematic is different, so the following steps only show an example of how reading a schematic works.

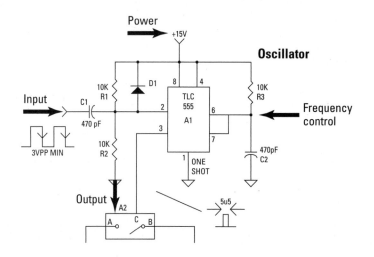

4. For each group, identify input and output signals for each and where they make connections in the circuit.

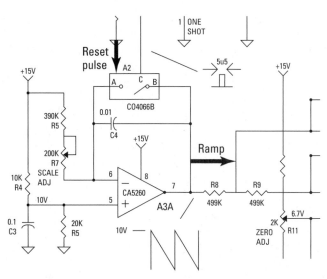

5. Trace the flow of important signals through the schematic.

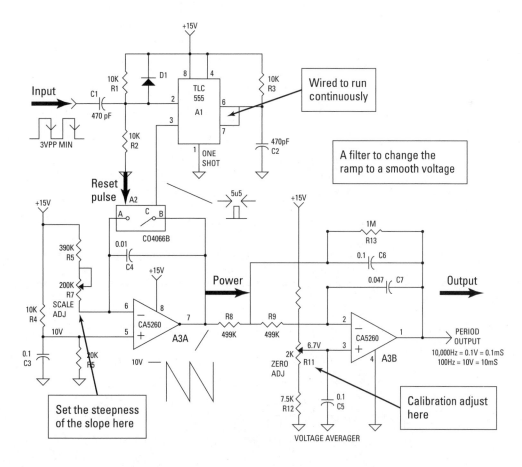

6. Identify and label the control signals and the internal connections for major functions and components.

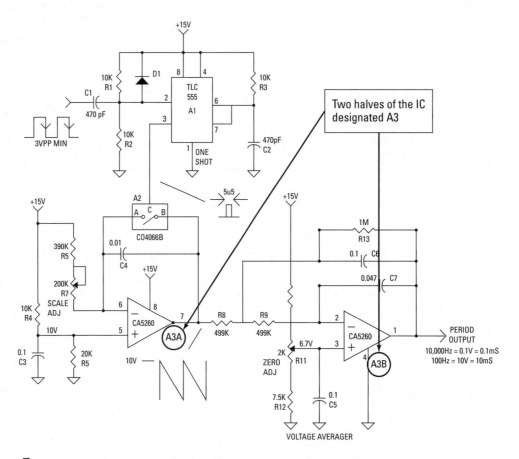

7. Some schematics include ICs that have multiple symbols, such as a quad NAND gate or dual op-amp, note where each part is located on the schematic. Identify any unfamiliar symbols and read any notes or legends on the drawing.

From here, you can dig in to understand the circuitry in each section. You may not understand every component's purpose right away, but try to grasp the general operation of each section. When you're finished, you'll have a pretty good feel for how the entire circuit is put together and how signals flow around the schematic.

Here are some additional resources for getting a handle on schematics:

- *Beginner's Guide to Reading Schematics* by Traister and Lisk.

- *Electronics For Dummies* by McComb and Boysen (Wiley Publishing, Inc.).

- *"How to Read Schematics"* an online tutorial at www.best-microcontroller-projects.com/how-to-read-schematics.html.

- *"How to Read A Schematic Diagram"* a downloadable article at www.arrl.org/tis/info/pdf/8402019.pdf.

Drawing your own schematic

It helps a lot to imagine yourself as the schematic's eventual reader, because someday you will be! Keep in mind the three functions for schematics (listed earlier in this chapter). Include all the information needed to build a circuit. Arrange the circuitry so your reader will be able to grasp its flow and function. Be sure that every important signal and connection is identified and labeled.

Here are some easy guidelines for creating legible schematics.

✔ Signal flow from input to output should be shown from left to right. (This isn't always possible in digital circuits.)

✔ Locate external connections at the sides of the drawing.

✔ Components that perform a common function should be placed near each other.

✔ Label component groups and important signals.

✔ If practical, include waveforms or voltages for normal operating conditions.

✔ Identify and date your schematic. Keep track of revision numbers to avoid later confusion between versions.

✔ Use standard symbols — and never use different symbols to represent the same component. For example, unless your circuit has more than one type of ground, use one ground symbol consistently.

✔ If there are special connection or construction requirements, show them clearly and include explanatory notes or references.

✔ Each component should have a type designator (R, C, L, IC, J, T, and so on) and identifying number (1, 2, 3, 4, and so on).

✔ Analog circuit schematics should have the most positive power-supply voltage at the top and the most negative at the bottom.

✔ On digital schematics, use the thick lines called *busses* for multiple parallel signals such as data and address signals. A small slash across the line with a number shows how many signals are combined, such as "8" for an eight-bit data bus or "16" for a 16-bit address bus.

✔ For complex ICs, group the connections by function instead of by pin number.

After you've finished the first draft, have a friend review it to see whether he or she can understand it too. Also, be sure to include a copy of the final schematic in your notebook with any construction and operating notes!

Part II
Building Circuits

In this part . . .

This is the part where you start building real circuits! There are many types and styles of circuitbuilding. Some are best suited for temporary experimentation; others enable you to make permanent circuits for the ages. This part helps you decide which approach is best for the job at hand.

You begin with a method called breadboarding that uses a special panel of contacts to hold components together. This convenient approach makes trying out a new circuit easy and inexpensive. You can make it permanent later!

Printed-circuit (PC) boards are by far the most common way of constructing a permanent circuit. This part introduces you to each of the two common styles of printed-circuit boards and gives you practice with handling the different components for each by guiding you through building a kit.

Three simple prototype projects show how easy it is to just wire something up to test out an idea or make up a one-time gadget. Then you get to try three of the most common hobby construction techniques: dead-bug style, Manhattan-style, and wire-wrap. Life at the workbench will never be the same, so heat up your soldering iron and get ready to build!

Chapter 3

Using a Solderless Breadboard

Tasks performed in this chapter

✔ Preparing your breadboard

✔ Building an audio amplifier

✔ Building a digital timer

The solderless breadboard — introduced in Chapter 1 — is a quick and easy way to experiment with low-power electronic circuits. Learning how to make use of a breadboard will save you countless hours of "workbench time" as you experiment and develop your own circuits and projects. In this chapter, you build a pair of simple circuits to get you familiar with using a breadboard.

Before you begin building a circuit on a solderless breadboard, there are a few simple techniques to keep in mind to minimize trouble:

✔ **Keep connections short.** As purchased, the leads of components such as resistors and capacitors are so long that the component is often left sticking up in the air above the breadboard once the leads are inserted into the contacts. In a densely-packed circuit, this allows the leads to flop around and short out adjacent components. A forest of leads also makes it difficult to see what is connected to what. If your circuit will have a lot of components, trim the leads so the components are close to — but not jammed against — the breadboard. The same goes for jumpers — keep them near the breadboard and bend them into shapes that make it easy to see where they're connected. It's easier to keep things organized if you keep a good selection of wire lengths handy.

✔ **Don't make those connections TOO short.** When stripping insulation off the jumpers, leave enough wire exposed so that when they're fully inserted, you can see some wire going into the hole of the plastic strip. If there isn't enough bare wire, it might *look* like the wire is connected when it really isn't. For component leads, make sure the lead can be fully inserted into the contact and firmly gripped. If you cut the lead too short, the component can easily pop out of the hole and not make contact.

✔ **Use color to your advantage.** Maybe you got a good deal on a zillion feet of wire with black insulation, but don't use it for *every* connection. You'd go crazy trying to keep track of the connections, and troubleshooting would be much more difficult than if you make consistent use of several colors. Any color convention will work, as long as you use it *consistently*. For example, surplus or used telephone home-wiring cable with four solid wires (the stiff, round cable, not the flat, flexible "modular" cable) is a great source of suitable (and cheap) wire that comes with a great color code built in, as follows:

- Black for ground

- Red for power supplies

- Yellow for control connections

- Green for signals or data

✔ **Track your progress.** For circuits with more than one complex integrated circuit (IC) or more than a dozen components, keep track of what you've wired up. Make a copy of the schematic diagram and *as you connect each wire or lead,* use a highlighter or colored pencil to trace the connection on the schematic. This is a good way to avoid missing a connection.

✔ **Make test points.** Use short pieces of bare wire to make convenient test points for measurements or for connecting to external cables. By providing dedicated connection points, you avoid having test leads pull component leads out of contacts, accidentally short out, or slip off an IC pin.

Ready to start building? Let's get going!

Breadboarding an Audio Amplifier

An audio amplifier is a very useful circuit and a good one for breadboard-building practice. This particular circuit uses an *integrated circuit* (IC) and several resistors and capacitors. If you pay attention to which pin is which on the IC — and double-check your wiring — chances are excellent that it will work properly the first time power is applied!

IC: Integrated circuit, also known as a *chip*. A collection of components, usually transistors, in a single package with multiple leads. Common examples of ICs include sets of digital logic gates, amplifiers, timers, and microprocessors.

Deciphering the amplifier schematic

Figure 3-1 shows the schematic of the amplifier circuit. (If you need help understanding schematics, refer to Chapter 2 of this book. The integrated circuit, IC1, is a type of amplifier called an *op-amp* (short for *operational amplifier*) because these amplifiers were originally used in analog computers to perform mathematical operations. The type of op-amp used here is Model 741.

The numbers surrounding the symbol are the pin numbers of the most common plastic DIP (dual in-line package) form of the 741. Pins 1 and 5 are left unconnected (nc for not connected). The signal flow is left to right from the input (V_{IN}) to the output (V_{OUT}). A single-polarity 12V DC power supply is assumed to be used, with the chassis ground symbol representing the ground or DC return terminal of the supply (also referred to as the negative terminal, even though the voltage is not negative with respect to ground). A battery or battery pack that can supply approximately 12 volts can also be used to power the circuit.

DIP: Dual In-line Package. The physical form of an IC that has a rectangular plastic body with one row of leads on each side in a line. DIP packages range from 6 to 64 leads.

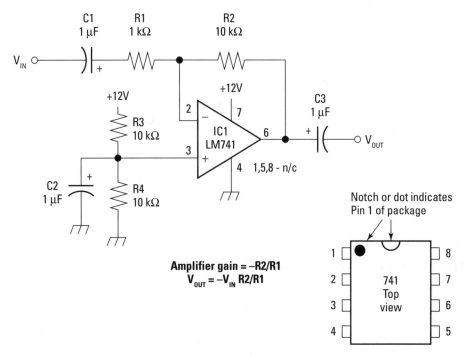

Figure 3-1: The schematic for a simple audio amplifier circuit using an op-amp IC.

How the audio amplifier works

Referring to the schematic in Figure 3-1, you can see the op-amp wired so current from the input (through C1 and R1) is balanced by an equal and opposite current through R2. (No current is allowed to flow *into* pin 2.) The voltage at pin 2 of the op-amp stays at the same voltage as pin 3. That's because R3 and R4 form a *voltage divider* (a resistor circuit whose output is a fraction of the applied voltage); they keep the DC voltage at pin 3 steady at 6V, which is about one half the power-supply voltage (see http://en.wikipedia.org/wiki/Voltage_divider for more). To balance the input current, the op-amp's output voltage must change to make the difference in voltage between pin 2 and pin 6 cause a balancing current to flow in R2.

If R1 and R2 are the same value, the output pin of the op-amp only has to have an equal-and-opposite voltage to V_{IN}. That means the *gain* of the amplifer — the factor by which the input signal is amplified — is –1. This is called an *inverting amplifier*.

If R1 and R2 have different values, then the op-amp has to swing more (if R2 > R1) or less (if R2 < R1) than the input. The gain of the amplifier circuit, A_V, is then the ratio of R2 to R1, multiplied by –1 because of the equal-but-opposite balancing rule:

$$A_V = \frac{V_{OUT}}{V_{IN}} = -\frac{R2}{R1}$$

Or

$$V_{OUT} = -V_{IN}\frac{R2}{R1}$$

Since R1 = 1kΩ and R2 = 10kΩ, $A_V = -10$

The function of capacitors C1 and C3 in the circuit is to block the DC voltage that will be present at pins 2 and 6 because a single supply polarity (+12V) is used. R3 and R4 cause the value of the output, pin 6, to be 6V when no input audio signal is applied. This is called *bias* and the 6V resting value is the *quiescent value* of output voltage. C2 filters the voltage at pin 3 and makes sure it remains a steady DC value by shunting any AC signal present to ground.

Breadboarding a Digital Timer

Solderless breadboards are very well suited to building *digital circuits* — those that use discrete voltage levels as signals representing logic values of TRUE and FALSE. With the amazing functionality of today's integrated circuits, even a small breadboard can be used to construct a sophisticated circuit.

To illustrate how to build digital circuits on a breadboard, we'll construct a digital timer circuit based on the MC14536B IC. This flexible IC consists of an *oscillator* (a circuit that generates a continuous stream of pulses), 24 *flip-flops* (circuits that divide the frequency of a stream of pulses by two), and some control inputs that configure or set up the IC to act in different ways. We are going to configure the control inputs so the IC counts a selectable number of pulses, turns on an LED (light-emitting diode), and then stops.

Digital timer schematic

Figure 3-2 shows the schematic for the digital timer circuit. The positive power supply voltage (+12V) is connected to the IC's power pin and several of the control signal lines and components at the top of the page. Ground is at the bottom, connected to the IC's ground pin and other IC inputs and components. Control inputs to the IC are shown on the left of the IC symbol and the output from the circuit is shown on the lower right.

Timer: In the world of digital circuits, a timer is any circuit that counts a fixed number of pulses before taking some action.

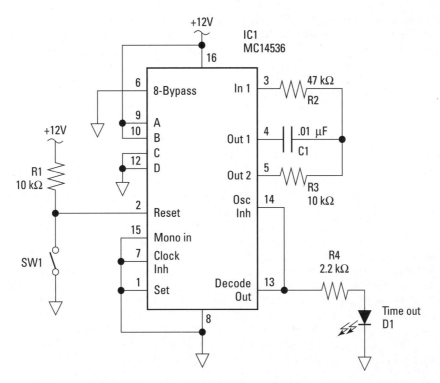

Figure 3-2: The schematic for the digital timer shows the IC in the middle with control inputs on the left and the output in the lower right-hand corner.

Why not draw a digital schematic with the pins organized around the ICs in the same order as the physical device? For the IC in the current circuit, that would place pin 1 in the upper left, pin 8 in the lower left, and so forth. While that the resulting schematic might have a stronger resemblance to the physical layout of the circuit, the circuit's functions would be obscured as the signal lines crossed each other and made numerous bends. Grouping signals and pins of like functions together makes the schematic and circuit much easier to understand.

How a digital timer works

Before I explain how the circuit works, download the IC's data sheet (as described in the tip at the start of this task). There is a lot of detailed engineering information you can just skip over as I explain how the timer works. Remember that if a 1 appears in the data sheet's truth tables and circuit descriptions, it means that a pin is connected to +12V (or puts out that voltage); a 0 means the pin is connected to ground or outputs a voltage close to 0V.

The basic function of the timer as we've configured it is to wait for a certain period by counting a fixed number of pulses and then turn on the LED. (the data sheet shows this schematic as "Time Interval Configuration Using the On-Chip Oscillator") Counting begins when the RESET signal (pin 2) is grounded (which happens when you close switch SW1). When the required number of pulses have been counted, the timer turns on the LED and stops counting. Opening the switch returns RESET to +12V (a logic 1) and that resets the timer. Closing the switch starts the cycle again.

The timer's oscillator circuit generates the pulses with a frequency determined by the value of C1 and R3 according to the formula:

Frequency in Hertz = 1 / (2.3 * C1 * R3)

The pulses are applied to a string of 24 consecutive flip-flops. Each flip-flop divides its input frequency by two, so the frequency gets lower and lower with each successive division. By the time the pulses have progressed all the way to flip-flip #24, the frequency of the pulses from the oscillator has been divided by 2^{24} or 1,677,216. If the frequency of the oscillator was 1,677,216 Hz (pulses per second), the output of flip-flop #24 would be 1 Hz (pulse per second).

Control signals A, B, C, and D (pins 9 through 12) select which one of the 24 flip-flop output signals is connected to the Decode Output (pin 13). Only one can be connected at a time. The data sheet's truth tables show the flip-flop output selected by each combination. Use the truth table with 0's in the column labeled "8-BYPASS"; that's because in our circuit, the 8-BYPASS pin (pin 6, connected to ground) is a 0. The schematic in Figure 3-2 shows ABCD equal to 1100 and that selects stage 12. The higher the number of the stage connected, the longer the time period from switch closing until the LED comes on.

The values for C1 and R3 result in an oscillator pulse frequency of approximately 4350 Hz. The first flip-flop output that can be selected (with ABCD = 0000) is flip-flop #9. Its output has been divided by 2 nine times or by 2^9, a division of 512. That frequency is 4350/512 = 8.5 Hz. So if I grounded A, B, C, and D — and then closed the switch — the LED would come on 1/8.5 = 0.12 second later. That's pretty fast!

By changing the connection of the ABCD inputs from 0 to 1 in various combinations, flip-flop outputs farther down the line will be selected. Selecting flip-flop output #12 (ABCD = 1100) results in a division by 2^{12} = 4096 or 1.06 Hz and the LED would wait for 0.94 seconds before coming on after the switch is closed. If ABCD were set to 1111 (all four connected to +12V) for the maximum division of 2^{24}, the output frequency would be a measly 0.004 Hz — you'd have to wait 241 seconds to see the LED come on after the switch was closed!

Constructing the Audio Amplifier

This section is a sequence of steps to get you from a handful of parts and an unpopulated breadboard to a working circuit ready for testing. Be sure you understand the instructions completely before starting each step. (Refer to the tools and techniques in Chapters 1 and 2 if you're unclear.) Don't forget to make a copy of the schematic for keeping track of your progress, as described at the beginning of this chapter.

1. Insert the leads (the wire pins coming out of the IC) of the LM741 op-amp IC into the breadboard, roughly in the middle of the plastic strip, straddling the center slot with each lead in a separate row of holes. Be sure that none of the eight leads are bent under the plastic package. Also be sure that two of the leads did not get stuffed into the same hole. (A magnifying-glass inspection is a good idea while you're learning.) Identify pin 1 of the IC.

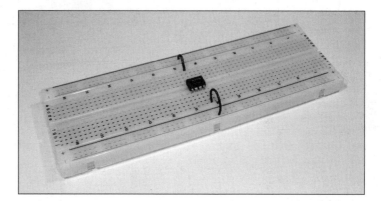

2. Decide which of the breadboard rails will be +12V and which will be ground. Install a short jumper between pin 7 of the IC and the +12V rail. Install another short jumper between pin 4 of the IC and the ground rail. Mark off these power connections on the schematic by tracing them with a highlighter or colored pencil.

3. Trim the leads on R3 and R4 to about 1 inch long, and then bend about half of each remaining lead at a right angle in the same direction so that the resistor looks like a "U" (as shown in Step 2's figure). Create the voltage divider by installing R3 and R4 near the IC so one lead from each resistor is in the same row of holes (connected together) and the remaining leads are in different rows of holes as shown in the Step 4 figure.

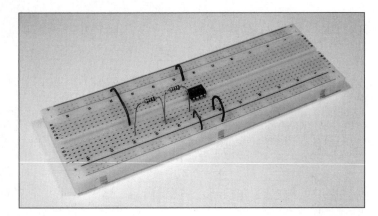

4. Add a jumper from the open lead of R3 (the one in a row all by itself) to the +12V rail. Then add another jumper from the open lead of R4 to the ground rail. Mark off both resistors, and then the power and ground connections, on the schematic. Don't mark off the line going to pin 3 of the IC or the connection to C2 — you haven't done those yet!

5. Complete the voltage divider with a jumper from the connection of R3 and R4 to pin 3 of the IC. Take a close look at C2 and carefully identify which lead is positive. Connect the positive lead of the capacitor to the connection between R3 and R4. Connect the negative lead of C2 to the ground rail or to an unused row of holes and add a jumper from there to the ground rail. Mark off both C2 connections and the connection from the divider to pin 3 of the IC. (Remember to track your progress on the schematic from here on out — good luck!)

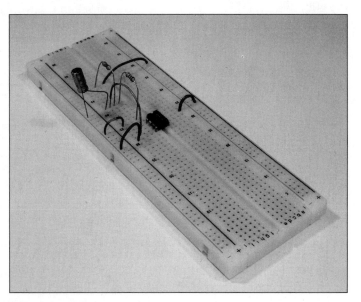

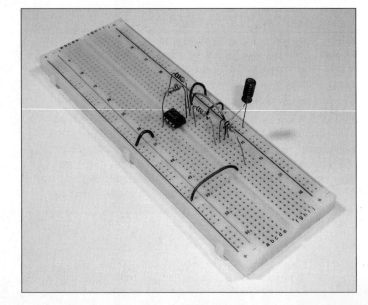

6. Bend the leads of gain-control resistor R2 so it can be connected directly between the rows of holes connected to pin 6 of the IC and pin 2 of the IC. It may stick up in the air a little bit, but that's okay since there aren't any other components around it. Connect one lead of the other gain-control resistor R1 to the row of holes connecting R2 and pin 2 of the IC. The remaining lead of R1 should be inserted in an unused row of holes nearby.

Electrolytic capacitors usually identify the negative lead with a minus sign in a round circle or an arrow pointing to the lead. Some may have a plus sign near the positive lead.

From here on, *connect* means "put a component lead into a hole" — got it? You bet!

7. After carefully identifying the positive lead of each capacitor, connect C1 from the open lead of R1 to an unused row of holes nearby. Connect C3 from the connection between R2 and pin 6 of the IC to another unused row of holes.

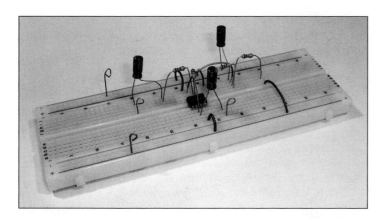

8. Create connection and test points by taking a ¾-to-1-inch piece of bare wire, making a small loop at one end, and inserting the straight end into the same row of holes as the open lead of C1. This is the *input-signal connection*. Do the same at the open lead of C3. This is the *output-signal connection*. Add two more test points on the ground and +12V rails. These are the *ground connections* for input and output signals. The small loops make good connection points for a test probe or clip.

Congratulations! You just built an amplifier circuit! All the lines on the schematic should now be colored, signifying that all connections are complete. Before applying voltage, carefully inspect the circuit (use your magnifier) to ensure that all connections are where you thought they were — and that all components and jumpers are fully inserted into their contacts. Be sure that none of the component leads are leaning against other leads.

If the capacitors are connected backwards with the negative lead more positive than the positive lead, they may be damaged when power is applied. When this happens, you may smell something pungent, the capacitor will probably get warm or even hot, it may emit what looks like smoke or fumes, and on occasion may give off a loud SNAP! This isn't a catastrophe — you just received an inexpensive lesson on what a bad electrolytic capacitor smells, looks, and sounds like! Nevertheless, use pliers to lift the afflicted component from the board and toss it in the garbage can. Wipe off any deposit from the breadboard and replace the capacitor, getting the polarity right this time.

Testing the Audio Amplifier Circuit

You've double-checked the circuit against the schematic. You're sure the IC and the capacitors are all installed with the correct orientation of their leads. You're ready to apply the juice and give it the smoke test electronic shop-talk for a circuit's first application of power.

The following steps assume that you have used a voltmeter before. If you're unfamiliar with how to use a voltmeter or connect a signal source to a circuit, refer to Part IV "Measuring and Testing" first.

1. If you have an adjustable power supply, be sure its output voltage is set to +12V.

2. Connect the rails to the power supply or battery pack terminals using clip leads or suitable wires.

3. Set your meter to read DC volts. Connect its negative probe to the ground test point, either with a clip lead or by using the probe itself (you may have to hold it in place). Connect the positive meter lead to the +12V test point.

4. It's time to switch on the power supply. Your meter should read +12V. If not, turn the power supply off immediately and correct any wiring problems. If it reads 0, one of the power supply connections is open or a connection to one of the rails is faulty or what you thought was a rail isn't a rail. Figure out which is which and try again. If the meter read −12V, the power supply or rail connections are backward. The IC is probably damaged if it is hot or even warm and may have to be replaced.

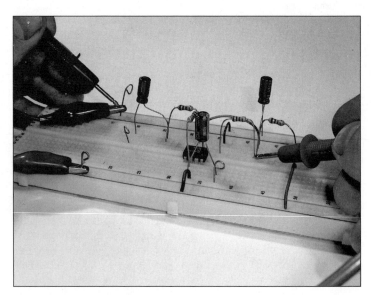

5. If the meter reads +12V, move the positive probe to pin 3 of the IC or any lead connected to pin 3 of the IC. It should read +6V. Pin 6 of the IC should also read +6V. If not, check the voltage divider wiring. There should be +6V at the connection between R3 and R4 and at pin 3 of the IC.

6. To apply AC voltage, set the output of your AC signal source (a signal or function generator or even a music player) somewhere between 0.1V and 0.5V when measured with a voltmeter set to read AC volts. (The exact level isn't critical.) Connect the ground lead of the audio source to the ground-rail connection point, and the hot lead of the audio source to the input-signal connection point. (If you need help making an audio cable or identifying connector wiring, Part III is the place to go for more information.)

7. Measure the output voltage by connecting the voltmeter to the output-signal connection point. The voltage here should be approximately 10 times the voltage at the input. If you have a pair of headphones or a speaker, connect it with clip leads to the amplifier output.

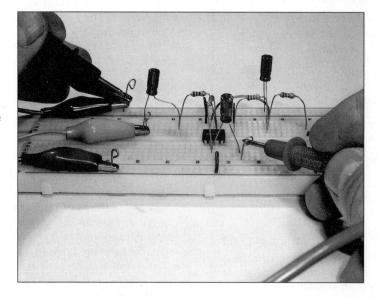

8. Don't stop now! Try some different resistor values for R1 and R2 to see what happens to the circuit gain. For example, swapping R1 and R2 (turn the power off first) will turn the gain from ×-10 to ×-one-tenth; the output will get smaller. Change R2 to 1 kΩ will make the gain equal to unity (same level in and out). With the output at a comfortable listening level, reduce the power-supply voltage if it is adjustable until the output signal begins to distort. Why does it distort? Because the amplifier output is being driven to the limit of the power supply voltage and can't increase any further — this is called *flattopping* or *clipping*.

Flattopping or **clipping:** When the input signal to an amplifier is too large for it to be reproduced, the circuit output voltage can not increase beyond the power supply voltage. If the output signal is observed with an oscilloscope, the portions of the signal that are too large to amplify are flat, straight lines and the waveform looks like its top has been "clipped off."

Constructing the Digital Timer

As in the previous task, this section is a sequence of steps to get you from a handful of parts and an unpopulated breadboard to a working circuit ready for testing. Be sure you understand the instructions completely before starting each step. Refer to the tools and techniques in Chapters 1 and 2 if you're unclear. Don't forget to make a copy of the schematic for keeping track of your progress as described at the beginning of this chapter!

Download the manufacturer's information for the IC (called a data sheet) by entering the IC part number and "data sheet" into an Internet search engine. I found the data sheet for the MC14536B at www.onsemi.com/pub/Collateral/MC14536B-D.PDF. The data sheets for many ICs and transistors can be found just as easily.

The MC14536B IC is sensitive to static. If you live in a dry climate (the kind in which you can generate a spark by walking across the carpet), you should take precautions to be sure you don't zap your IC by accident. Static dissipation mats are available from most electronic vendors, but an inexpensive substitute is to tape a strip of aluminum foil to the front of your workbench. Ground the strip by connecting it to a 10 kΩ resistor, using clip leads or wire taped to the aluminum foil. Connect the other lead of the resistor to the negative terminal of your power supply or battery pack. As you work on the circuit — and every time you approach the bench — be sure to touch the foil to discharge the static charge carried by your body before touching any circuitry.

1. Insert the leads of the MC14536 IC into the breadboard, roughly in the middle of the plastic strip, straddling the center slot with each lead in a separate row of holes. Be sure that none of the 16 leads are bent under the plastic package. Also be sure that two of the leads do not get stuffed in the same hole. Use a magnifying glass to take a close look. Identify pin 1 of the IC.

2. Decide which of the breadboard rails will be +12V and which will be ground. Install a short jumper between pin 16 of the IC and the +12V rail. Install another short jumper between pin 8 of the IC and the ground rail. Mark off these power connections on the schematic by tracing them with highlighter or colored pencil.

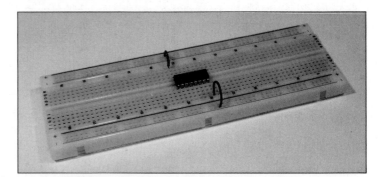

3. Note that several of the control-signal pins (1, 6, 7, and 15) are connected to ground. Use short wires to connect all these pins together. Use one more short wire to connect pins 7 and 8, grounding them all. Remember to mark off those connections on your schematic — and do so at every step.

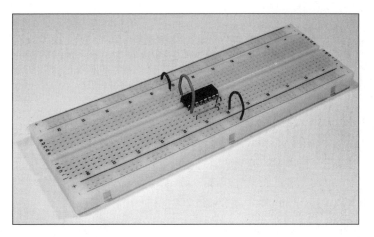

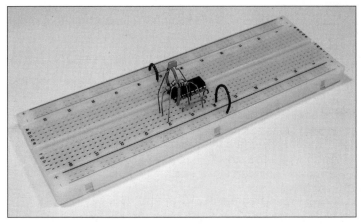

4. Wire the oscillator section by connecting one lead of R2 to pin 3. Connect one lead of C1 to pin 4. Connect one lead of R3 to pin 5. The remaining leads of these components should all be connected together by plugging them into an unused row of holes near the IC.

5. Wire the output section by connecting pins 13 and 14 together with a short wire. Connect one lead of R4 to pin 13. Connect the remaining lead of R4 to the anode lead of *light-emitting diode* (LED) D1. Connect the cathode lead of D1 to the ground rail.

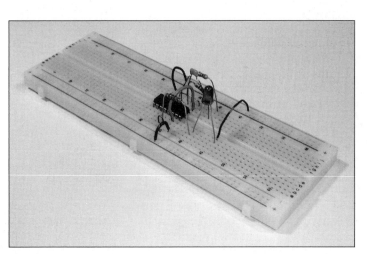

If you are unsure which lead of D1 is the anode, the packages in which LEDs are sold often feature a drawing showing the anode and cathode — and a round LED package usually has a flat side identifying the cathode lead. You can also test the diode with your multimeter set to measure resistance through the diode. With the meter showing some resistance (instead of an open circuit), the lead connected to the meter's positive (+) terminal is also connected to the LED's anode lead.

6. Connect the RESET switch pins to unused rows of connections. The pins of most miniature toggle switches cannot be directly inserted into breadboard contact holes — they are too big. Solder short pieces of bare wire to the switch pins to make breadboard-compatible pins.

7. Connect one of the switch's pins to ground and the other pin to pin 2 of the IC. (If your switch has three pins instead of two, use your multimeter to find the pair of pins on the switch that open and close when you move the switch handle from one position to another. Make a note of which position of the handle closes the contacts between that pair of pins; the timer will run with the switch in that position. Leave the third pin unconnected.) Connect one lead of R1 to pin 2 of the IC and the other lead to pin 16 of the IC.

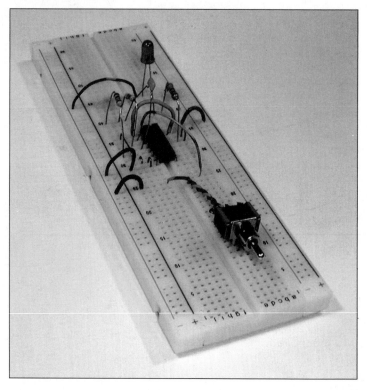

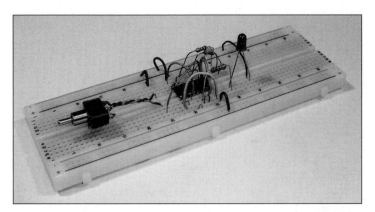

8. Wire the control signals A, B, C, and D as shown. Pins 9 and 10 should be connected to +12V and pins 11 and 12 to ground.

9. Make sure your power supply is set to +12V, open switch SW1, and apply power to the breadboard's +12V and ground rails. Use your multimeter to check the voltage between pin 16 and pin 8 — it should be +12 volts. Check each pin to see that it has the correct voltage; +12V or ground, according to the schematic. The clock-circuit pins (3, 4, and 5) will have some intermediate voltages — and that's fine.

10. Close the switch and verify that the LED comes on about 1 second afterward. The actual time may be anywhere from a half second to several seconds. The important thing is that the timer counts pulses and then turns on the LED.

11. Experiment with different combinations of A, B, C, and D to see what happens as you set the timer to different periods. (See the section "How a Digital Timer Works" earlier in this chapter.)

If your circuit doesn't work, go back over the schematic very carefully and double-check your wiring. Make sure leads are inserted into the holes you thought they were. Sometimes it helps to have a friend who hasn't worked on the circuit compare it to the schematic independently. Digital circuits generally work if connected properly, so the likely problem is a wiring error.

Chapter 4

Building a Printed Circuit Board

Nearly all electronic devices you can buy include one or more printed circuit boards (or PC boards). Luckily for the beginner, many kits are just right for learning how to use a PC board, like the learn-to-solder kit mentioned in Chapter 2.

There are two types of PC boards; *through-hole* and *surface-mount*. Components on a through-hole board have wire leads or terminals that extend through holes in the board, thus the name. Surface-mount boards, on the other hand, have no holes and the components are soldered directly to the copper cladding.

Surface-mount technology (or SMT) components are much smaller than *leaded* components with wire leads (pronounced "leeds"), allowing the PC boards and the equipment that uses them to be much smaller than if through-hole boards are used. SMT components have either metal caps or very short leads designed to lie flat on the surface of the board. To the hobbyist, though, the components are much smaller than leaded parts and are harder to work with using regular soldering tools.

Printed circuit board (PCB): A PCB is made from a core layer of insulating material such as fiberglass or plastic to which a thin sheet of copper (called *cladding*) has been attached on one or both sides. The desired pattern of conducting paths (called *traces*) is then printed on the copper. The PCB is then placed in a chemical solution that removes all of the copper except that covered by the printed pattern, leaving only the traces on the insulating core.

In this chapter, you'll learn how to work with each type of board by building a kit. You may choose to buy a different type of kit, but the techniques used are the same for all kits that use through-hole or surface-mount.

Getting Your Workspace Ready

Before beginning to build either type of kit, you need to prepare your workspace. The most important thing for all phases and types of circuitbuilding is good lighting. You'll be peering at the tip of your soldering iron and trying to read tiny part markings. Working with poor lighting is no fun — and it leads to mistakes. A swing-arm desk lamp is a good temporary solution if you don't have a permanent workbench or table.

Holding the PC board is also important. Trying to work on a board that slides around on the work surface can be frustrating — and can damage components or knock them out of alignment just as you're trying to solder them. A small machinist's vise will work for small boards. If you plan on building a lot of electronic stuff, you might want to invest in a Panavise (www.panavise.com).

As with any project, it's a good idea to keep your work area clean. Soon enough, you'll fill it with tools and bits of this and that. But when you're just learning, avoid distracting clutter in your workspace. If you are using a woodworking shop or maybe the kitchen table, buy a sheet of white poster board or cardboard to act as a work surface. You'll be able to see the parts much more clearly — and it will protect the underlying surface, too.

Putting a Through-Hole PC Board Together

If you enter the phrase "electronic kits" into an Internet search engine, you'll discover dozens of companies that sell hobby- and professional-level kits. Your author needed a battery charger, and so selected the Ramsey Electronics (www.ramsey-electronics.com) LABC1C Lead-Acid Battery Charger shown in Figure 4-1. You don't have to build this particular kit — choose one that fills some need in your shop or around your home. If this is your first kit, select one that uses mostly discrete or individual parts (such as resistors, capacitors, transistors, and diodes) rather than pre-assembled modules or integrated circuits. These parts are easy to work with and solder; if you get familiar with them, then when you graduate to more complicated kits, you'll be ready.

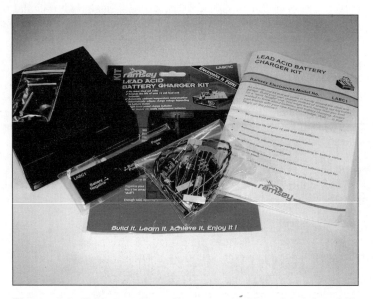

Figure 4-1: The battery charger kit uses a typical through-hole board. The components have either wire leads or terminals that stick through the board and are soldered to exposed pads on the bottom of the board.

Building a Surface-Mount PC Board

Surface-mount PC boards are relatively new, especially to hobbyists, and require different techniques than the familiar through-hole boards. To teach circuitbuilders how to use surface-mount components and PC boards, several companies make training kits.

Building one of these kits usually creates a simple gadget to have fun with, as well as teaching you the basics of working with surface-mount components. One good kit is the SM-200K SMT Training Course kit from Elenco Electronics (www.elenco.com). The kit has a few practice components to get you started; when you're done, you'll have a silly "Decision Maker" gadget, as well! If you choose a different training kit, it will have similar steps and procedures.

Because the parts are so small, a head-mounted magnifier is very handy. These are available for a few dollars at fabric and craft stores. Single magnification of a few powers of magnification is fine — you do not need dual or high magnification.

You are also likely to drop a part or two. They bounce incredibly well! So well, in fact, that they are likely to wind up on the floor. A smooth, light-colored floor helps find the parts, but if that doesn't describe YOUR floor, spread out an old, light-colored sheet around the workplace so you can find that tiny part after it slips out of the tweezers.

Figure 4-2 shows the basic tools that you'll need to work on SMT assemblies. The stainless-steel tweezers on the right are a good investment if you plan to build or repair electronics that use surface-mount technology.

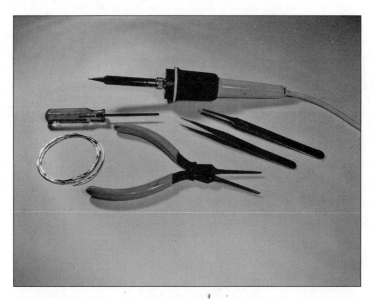

Figure 4-2: Surface-mount tools. Note the fine tip on the soldering iron.

Constructing the Through-Hole Board

The PC board for most hobby kits is at most a few inches on a side, and contains fewer than 100 components. The kit may come with an enclosure or it may just be a plain PC board that you can use by itself. In either case, the basic procedure shown in this task applies to both types of kits.

Before buying the kit, check to see if it requires a separate power supply, enclosure, or other materials that you have to provide.

The figures show how to construct most through-hold PC boards. The task will go well if you stay organized and follow the manufacturer's instructions carefully. Resist the temptation to do things out of order or skip steps. That's when mistakes are made — even if you have a little experience already!

As you build the kit, take notes (remember the notebook!) about construction and design details. You can apply these in your own designs later!

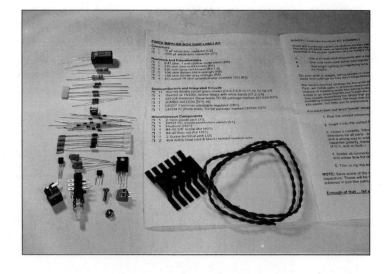

1. Open the kit and find the instructions and the parts list. Check off the parts and pieces against the parts list to be sure you have everything. It's also a good way to familiarize yourself with all the components.

> **TIP**
> Use the kit's instructions — especially the checklist — to help keep you from missing information or skipping steps by accident. Experienced circuitbuilders have learned (usually the hard way) that the manufacturer usually gives the instructions in a particular order for good reason!

2. Sort the parts by type into a container that is divided into bins or sections. An inexpensive muffin or cupcake baking tray is popular in the workshop. Or look around your home — the cardboard bins in the figure were used as packing material! Egg cartons work well, too! You can also stick the leads of parts into a scrap piece of Styrofoam covered with aluminum foil (the foil eliminates static) to hold them while you work.

3. Read the assembly instructions thoroughly, even though you're anxious to get started. Build a picture in your mind of what you're going to do, and in what order. Look for sequences of steps that must be done in a specific order — and make note of warnings and cautions.

Look especially hard for errata in the manual or on slips of paper included with the parts. It's not uncommon for companies to have to substitute parts — or to discover and correct a layout or manual error. If you do find errata note them in the instruction manual or attach the errata sheet to the manual on the corresponding page so you don't forget about it.

4. As you insert leaded parts (diodes, resistors, capacitors, inductors) into the holes, hold them to the PC board by bending the leads slightly apart. Bend them over at the PCB surface just enough so they stay in the holes and don't slide back out when you flip the board over to solder them. Don't bend the leads so much that they lie too close to the board (which would make them difficult to solder and trim).

If you are attaching an integrated circuit (IC) or an IC socket to the board, bend the leads at opposite corners of the rows of pins to hold the part to the board. Solder those leads first.

Leads

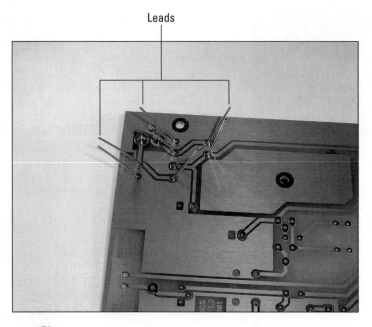

5. Solder the leads to the PC board pads. Don't use too much solder — you just want to cover the pad and flow solder onto the component lead. For components with short, thick leads that don't bend easily, melt a small ball of solder onto the iron tip, and then tack-solder one or two leads to hold the part. Then solder the rest of the leads, and go back to finish the job on the tack-soldered leads.

Be sure components such as controls and switches are fully inserted when you're soldering them; mechanical alignment may be important for the panel or enclosure mounting holes.

You might wonder why some single-sided PC board kits use jumper wires instead of routing all of the traces on the cladding side. The reason is simple: single-sided PC board is cheaper than double-sided!

6. Do a close visual inspection of solder joints before you move on to the next step. Look for the smooth shiny surface that shows the solder has flowed evenly over both the PC board pad and component lead before cooling and solidifying. Trim the leads just beyond where the solder joint wets the lead. Save the trimmed bits of wire for jumpers or other uses.

7. When you are installing polarized components such as capacitors and diodes, pay particular attention to polarity markings.

Heat sink   Regulator IC

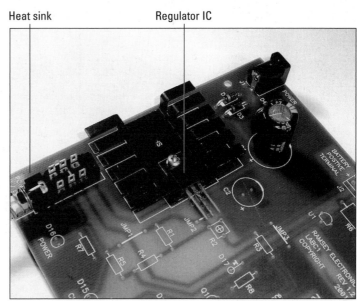

8. Some kits have mechanical assemblies such as the heat sink and regulator IC shown here. Make a dry run of any mechanical assemblies to be sure everything lines up where you expect. If it doesn't look right, don't complete the assembly until you've figured it out.

REMEMBER

Tighten up the mechanical assembly *before* soldering the leads. Soldering last minimizes mechanical stress on the solder joints.

Transistor

9. When installing transistors, don't force them down against the board. Spread their leads a little bit and insert until the leads feel tight in the holes.

10. Complete the checklist and give everything a close visual inspection — look for polarity, orientation, values, shorts, solder balls and solder bridges between pins or terminals. Brush the back side of the board with a stiff brush to clear off debris.

11. Install the board in the enclosure, if there is one. Perform the test and alignment procedures in the manufacturer's instructions. Congratulations!

You might have noticed that in the parts list step, both LEDs were the same color. The preceding step shows two *different*-color LEDs. What happened? Well, the author customized a little bit. If the LEDs are different colors (the one for power is green and so looks grayish in the figure and one that says full charge has been reached is red and appears dark), it's easier to look at the charger from across the room and check its status. Feel free to customize the kit (it's *your* kit, after all) as long as the changes don't affect its function. Substitute different color LEDs, change a connector, or even add a connector or switch. Don't be afraid to experiment!

Constructing the Surface-Mount Board

Stuff You Need to Know

Toolbox:
- Fine needle-nose pliers, wire cutters
- Fine-tip soldering iron and solder (thinner than 1 mm)
- Small flat-blade screw-driver or toothpick
- Magnifying glass or head-mounted magnifier
- Tweezers
- Component holding tray

Materials:
- Surface-mount PC board training kit

Time Needed:
About half a day

If you have constructed a through-hole PC board, then you already know how to do most of the necessary steps to construct surface-mount boards. What will be new is the scale at which you work.

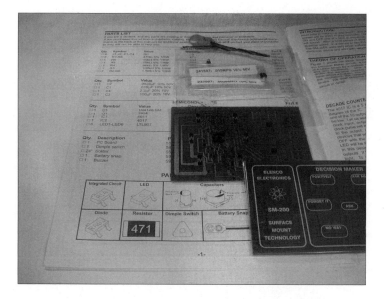

1. Open the kit and find the instructions and the parts list. Check off the parts and pieces against the parts list to be sure you received all of them. (Use a magnifying glass to read the tiny part markings.) SMT components are very small, so take your time and be careful. To pick components out of tape strips, pick the plastic film off with a sharp knife or tweezers.

2. Read through the instructions carefully. Take a few moments to familiarize with this type of component and the techniques for assembling a surface-mount PC board.

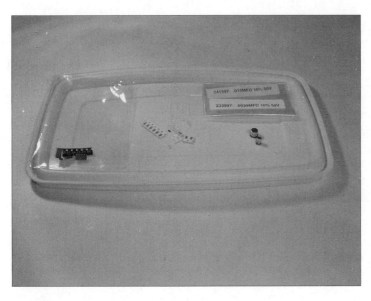

3. Instead of a set of bins for parts, a refrigerator storage container lid will work just as well. Keep the parts in bags by type if there are a lot of parts.

Keep any ICs in their shipping bags due to static sensitivity!

4. Most training kits have some practice components and pads on the PC board. This feature allows you to get used to the scale of soldering SMT parts before beginning the assembly for real. Don't proceed until you have learned the basic soldering steps. Your first couple of attempts might be pretty lame, but that happens to everybody.

5. You will soon develop your own style of SMT soldering. The author prefers to begin by lightly tinning the pad. Using tweezers, place the component on the tinned pads. Put a very small ball of solder on the tip of the iron. Hold the component against the board with the tip of the tweezers. Touch the iron to the component lead and pad simultaneously to flow the solder over both.

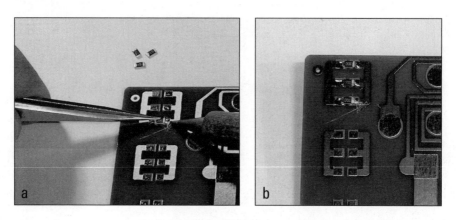

a

b

6. Start with the resistors and capacitors. Take your time and inspect each component with a magnifier after soldering. Look for a smooth fillet between each pad and the end cap of the component you're attaching.

TIP

As you work on the board, turn it to get the easiest access to the component. Soldering SMT parts is tough enough by hand without having to lean over the board awkwardly.

WARNING

If you get tired, STOP! This is when you make mistakes, slip with the soldering iron, or knock parts on the floor. And you get frustrated. At least take a break, get a little fresh air, and be ready to have fun when you come back to the project. It's not a race!

7. For multilead packages (transistors, diodes, and ICs) tack-solder one lead first, then adjust the part's position until it's in exactly the right spot.

8. To solder the large capacitors, start by tinning the pads. Position the capacitor, then heat one pad while lightly pressing the capacitor down onto the board. Release the capacitor when the solder melts and the lead contacts the pad. (You'll be able to feel it.) Check alignment, then solder the other pad.

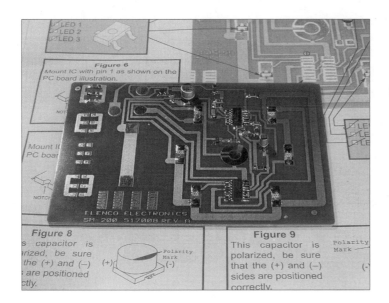

9. When you're soldering ICs, use the tack-solder technique in Step 7 to hold the IC on the board. Use the magnifier to make sure every lead is approximately centered on its pad. Hold the iron at the far end of the pad, touch it with the solder, and then bring the molten solder up to the IC lead by dragging the tip of the iron along the pad.

Chapter 5

Building a Prototype

Tasks performed in this chapter

✔ Assembling an audio level controller

✔ Building a voltage regulator

✔ Adapting a kit to create an alarm

Most home and hobby electronics projects consist of *prototyping*. "Proto" means "first," so a *prototype* is the first version, model, or instance of something being built. All big projects and products begin with a prototype, just to see how the idea works out. Since most of us are not in business to construct products, just about everything we build can be considered a prototype! If you go on to improve your circuit, as many of us do, each model is a prototype for the next.

Much of the work of prototyping involves packaging the electronics. You may build the electronics from individual components — some techniques for building this type of circuit are covered in Chapter 6. As electronics manufacturing has become more and more efficient, it has also become more and more common (and cheaper!) to simply use ICs or modules that do the job. All you have to do is provide power, input and output connections, and the necessary controls. That's what the tasks in this chapter will illustrate.

It's quite unlikely that all of the parts you have or buy will exactly match those in this book. Don't be afraid to substitute and experiment! That's the fun of electronics as a hobby — trying something new and learning. If a figure shows a pushbutton switch and you have a toggle switch, that's fine! If you'd rather build your prototype in a coffee can than a plastic box, have at it. It's *your* prototype! The goal of this chapter is to give you some ideas about constructing electronic equipment and give you the confidence that you can create useful and interesting stuff at home.

While you're at it, don't forget to document what you build so that new technique or circuit is recorded for posterity! Keep a notebook handy for notes, schematics, layouts, plans, and ideas. You won't regret it when you need to duplicate or repair one of your creations.

Building an Audio Level Controller

This audio level controller is a good example of the type of simple gadget that a beginning circuitbuilder needs around the shop. It's quite common to want to use the audio output of a music player or stereo as input to some other circuit or recorder. However, the output voltage of these devices is often too high to apply directly to a recording input. Some kind of level reducer is needed.

Tapered pots

Tapered pots are not something you find in the garden department of the hardware store! If you look through an electronic parts catalog, pots are shown as having a *taper*. Taper is the way in which the resistance from the wiper to the ends of the element changes as the wiper is moved. For a *linear taper* pot, the wiper-to-element end resistances are proportional to the position of the wiper. For example, If the wiper is 25 percent of the way along the element, the resistances between the wiper and the end of the elements will be 25% and 75% of the full element resistance. Dial markings showing equal changes in wiper-to-element resistance will be equally spaced around the pot.

The resistance of *logarithmic taper* (or *log taper*) pots is weighted towards one end of the element so the logarithm of the resistance changes equally along the element. This is useful in certain kinds of amplifier and control circuits.

In a linear taper pot used as a volume control, most of the volume change occurs before the wiper moves halfway along the element. *Audio taper* pots are similar to log taper pots, but the change in resistance varies along the element; when the pot is used as a volume control, equal changes in wiper position result in equal changes in volume to the human ear.

Figure 5-1 shows the schematic of a simple level controller. The left and right audio signals are each connected to a variable resistor (also called a *potentiometer* or just *pot* for short). The *wiper* of the pot (denoted by the arrow on the schematic symbol) slides along the surface of a resistive *element* (the resistor symbol). The position of the wiper on the element determines the resistance from the wiper to each end of the element. With one end of the element grounded and the signal applied to the other end, the wiper forms a *voltage divider*. The voltage on the wiper gets smaller as the wiper moves toward the grounded end of the element.

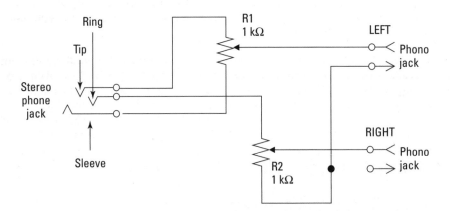

Figure 5-1: The Audio level controller: a pair of variable resistors (potentiometers) divide voltage and reduce input signal level.

The total resistance of the pots in the controller is 1 kΩ. If the amount of resistance from the signal end of the element to the wiper is R_A and the resistance between the wiper and ground R_B, then $R_A + R_B = 1$ kΩ. The output signal level from the wiper will be as follows:

$$V_{OUT} = V_{IN} \times \frac{R_B}{(R_A + R_B)} = V_{IN} \times \frac{R_B}{1k\Omega}$$

As the wiper moves along the element from the end at which the signal is applied toward ground, the output voltage gets smaller and smaller. The output signal can be varied from nearly the full input level to nearly nothing. This allows almost any audio source to be used for the input signal. If a *splitter* (an adapter used to allow two devices to connect to a single audio source) is used at the audio source's output, the recording device can even be operated while you listen to the music at full volume!

When you select an enclosure for your controller, make sure it's big enough to hold all the components — it's easy to underestimate how much depth or width is required. The enclosure should be large enough that you can access the terminals of the connectors and pots. If the enclosure is big enough that all of the components can be mounted on the lid (as shown in the photos), even better!

For more information on phone and phono-type headphone and stereo connectors, refer to the information about making cables in Chapter 10. You'll need a stereo patch cable to connect the level controller to the output of the audio source. In this chapter, the output of the controller is shown using separate phono jacks (typical of audio recording inputs). You may choose to substitute another phone jack if you intend to use the level controller with a computer sound-card input.

Building a 12V-to-5V Regulator

Making one of these regulators is a good example of using the power of an IC to perform almost all of the functions of a prototype. All the significant circuitry is contained within the single, three-lead body of a *voltage regulator* IC. Your job, as prototype designer, is to package it all so the circuitry stays cool and the input and output connections are secure.

The regulator changes the output of your car's electrical system (nominally 12V, but often anywhere between 10V and 15V) to a smooth, even 5V. The process of controlling an output voltage is called (logically enough) *regulation* and the devices that do the job are called *regulators*. The regulator will come in handy for running digital logic projects you might build for your car; it can also control devices that run from three 1.5V batteries, since 5V is reasonably close to the output voltage from three fresh batteries.

Building an Audible Alarm

In this task, you'll adapt a commercial kit to a new purpose — making a simple alarm. There are so many good kits available that sometimes it's easier to use one for a new purpose than to design a completely new circuit from scratch. In this case, using an oscillator kit that costs less than $10, adding a speaker and enclosure from the junk box, a capacitor and a connector, and you've got a perfectly good alarm!

An alarm is a particularly useful thing to build. It doesn't have to be designed to catch a burglar; it can respond to any event that can close an electrical switch. You can use a thermostat, an airflow switch, a light-dark detector, a pressure switch, and even (yes) a security alarm switch. You can find a simple schematic for doing so at the end of the task, later in this chapter.

You'll have to choose from one of the many available oscillator or pulse-generator kits available. The Ramsey Electronics (www.ramseyelectronics.com) Kit UT-5, the "Universal Timer," is based on the popular 555 timer IC. (See Chapter 3 for more information about the 555 timer.) Depending on your intended use, the circuit can be wired as a delay, a single-pulse generator, or it can generate a continuous stream of pulses. Any oscillator kit that can drive a speaker to satisfactory volumes will work in this simple system.

The continuous mode is also known as *astable* (that is, "not stable," because the circuit jumps back and forth between two states, OFF and ON). When the kit is configured for the astable mode (certain components and jumper wires are installed), the output is a stream of pulses whenever power is applied. The frequency of the pulses is adjustable by a variable resistor. The speaker is connected directly to the output of the oscillator. The 10μF *DC-blocking* capacitor is required to prevent the speaker from carrying DC currents from the kit's output. Figure 5-2 shows how the addition of the external components creates the alarm.

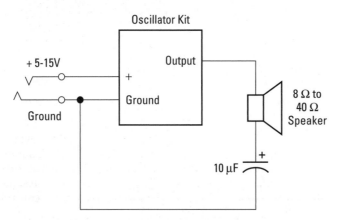

Figure 5-2: Adding a speaker to the output of the oscillator creates an audible alarm whenever an external circuit applies power to the oscillator.

Constructing the Level Controller

Before you begin building anything, make a photocopy of the schematic and make sure you have all the parts ready to go. You'll need to mark off each connection as it's made. Don't forget to keep a notebook handy for notes, schematics, layouts, plans, and ideas.

1. Lay out where each control and connector will go on the enclosure. Look carefully for mechanical interference or obstructions inside the enclosure that would prevent mounting a connector in the desired location. If some of the parts are mounted on the lid and some on the side, will they stay clear each other when the box is assembled? Mark the location of each connector and control.

TIP

If the hole is a mounting hole for a sheet metal or self-tapping screw, use a drill bit that is the same size as the *inside* of the threads. Hold the drill bit against the screw threads with the screw behind the bit. The threads should stick out past the sides of the bit, but the center portion of the screw should not be visible.

2. Centerpunch where each hole will be drilled.

3. Drill small pilot holes at each location with a ⅛" drill bit. (Any size close to that will do.) If the holes are to be used for mounting screws, use a drill size guide to select the proper bit size.

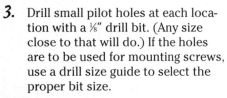

Plastic enclosures can be centerpunched with a scratch awl as shown in Step 2's figure. Hand drills also work well with plastics, and are less likely than a high-speed electric drill to wander off center and scratch the surface.

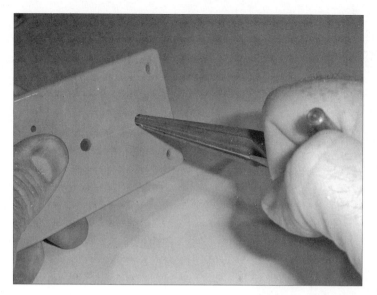

4. Use a hand reamer to increase the size of the holes if you're using a plastic enclosure. The hand reamer will gradually enlarge the hole without seizing in the hole or cracking the plastic. Reamers also work well in thin metal enclosures; drill bits tend to grab and tear sheet metal. If you decide to drill out the holes, use a vise to hold the material.

5. Mount all of the connectors and controls on the enclosure. Use lock washers under nuts inside the enclosure. Flat washers should be used on the enclosure's outside surface for any of the controls or connectors that mount in a single hole, such as the phone jack and volume control shown in the figures. Tin all contacts and terminals.

 When using a reamer, turn it continuously in one direction. This helps ensure that the hole will be round and not oblong or star-shaped.

When connectors where the terminals are very close together, such as output phono plugs, it may be easier to drill small holes for each and then use a file or utility knife to make a slot-shaped hole.

6. Use a multimeter's resistance scale or continuity tester to determine which of the phone jack terminals are Tip, Ring, and Sleeve. The Tip contact will be the one that contacts the plug farthest from the mounting bushing. Insert a phone plug into the jack if there is any doubt about which contact is which.

7. Use the multimeter's resistance scale to identify the wiper terminals of the pots. (They're usually the ones in the center of the three terminals.) Now turn the shaft of the control all the way counterclockwise. This will be the lowest-level setting. Determine which of the remaining two terminals has the lowest resistance to the wiper terminal. That terminal will be the one connected to ground; the other will be connected to the input signal.

Bare wires

8. Connect the phone jack's sleeve terminal to the pot's ground terminals and the phono jack's shell terminals. You can make individual connections with separate pieces of wire or run a single bare wire to connect all five terminals (as shown in the figure).

9. Measure and cut pieces of wire to connect the signal terminals. The wires should be long enough that they're not under any tension when soldered to the terminals. Strip and tin each end of the wires. Solder one end of each wire to an input or output connection.

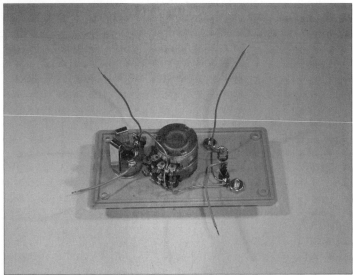

Get into the habit of using different colored wires for different audio channels. That makes tracing connections a lot easier and reduces the chances of error when wiring the circuit.

10. Connect the signal wires from the input connectors to the signal input terminals of the pot. Connect the signal wires from the output connectors to the wiper terminals of the pot.

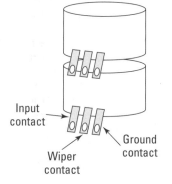

Input contact

Wiper contact

Ground contact

11. Assemble the enclosure and install the knob on the pot. To align the knob, turn the shaft of the pot fully counterclockwise and orient the knob so that it will be in the "7 o'clock" position. Tighten the knob setscrew(s). Label the box any way you choose — you're done!

12. Apply an input signal from a stereo audio source and connect the outputs to a stereo receiver or other device with separate channel inputs. While listening to the output of the controller, adjust the levels and be sure turning the knob clockwise increases the output levels and vice versa.

If the levels increase when the knob is turned counterclockwise, you have the ground and signal terminals of the pot reversed. Swap the ground and signal connections. If the channels themselves are reversed (left channel audio on the input appears as right channel audio on the output), exchange the signal wires at the input connector. (Don't change the ground connection.)

Constructing the Regulator

Stuff You Need to Know

Toolbox:

- ✔ Fine needle-nose pliers, wire cutters, soldering iron and solder, drill and reamer

Materials:

- ✔ Plastic enclosure (don't use metal for this task)
- ✔ PC board (single- or double-sided will do)
- ✔ 7805 voltage regulator IC
- ✔ 1N4001 rectifier diode
- ✔ 0.1µF ceramic capacitor
- ✔ 1µF tantalum 16V capacitor
- ✔ 6' cigarette lighter cable with or without an output connector

Time Needed:
Less than half a day

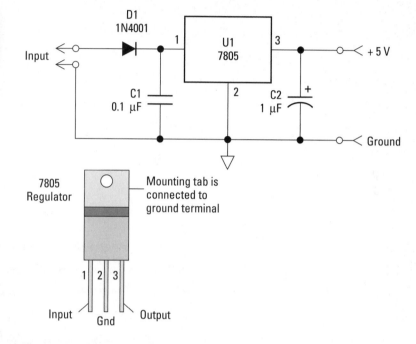

The regulator IC is a type of circuit known as a *linear regulator*. The term "linear" is used because the circuitry inside the regulator makes continuous adjustments to keep the output voltage at 5 V. Regulation of the output voltage is achieved by adjusting the drive current to an internal *pass transistor* through which all of the regulator output current flows. So the linear regulator is really just a kind of smart resistor. The circuitry acts like a resistor whose value constantly changes to just the right value, creating enough voltage drop from the input to the output to maintain a 5V output voltage. (For more information on voltage drops and Ohm's Law, see Chapter 12.)

Like a resistor, the regulator also generates heat. The amount of heat generated in watts (called *power dissipation*) is equal to the current through the regulator times the voltage from its input to output leads. For example, if 100 mA is flowing through the regulator with 12V at the input and 5V on the output, the power that must be dissipated is $0.1 \times (12 - 5) = 0.7$ watts. This amount of heat will cause the regulator IC to get warm to the touch, but not blistering hot. Raise the current close to 1 amp, however, and the entire circuit board will get warm as the heat is transferred through the regulator's metal tab to the PC board.

The 7805 regulator has another trick up its sleeve, however. If it senses its temperature rising too high, it shuts itself off until it cools back down! In that sense, it is almost immune to short-circuits! You can tell when the regulator is engaging in *thermal shutdown* because the output voltage will drop to a low value, hiccuping back to 5V for short intervals before being turned off again.

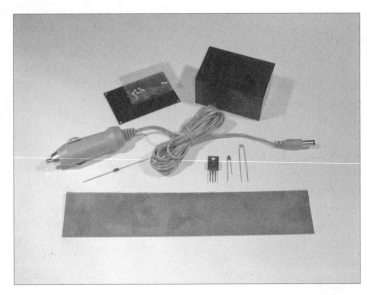

1. Start by making a photocopy of the schematic and making sure you have all the parts ready to go. You'll need to mark off each connection as it's made.

2. Cut a piece of PC board to fit the inside of your enclosure using a sharp utility knife and a straight-edge. Place the straightedge on the PC board and score it with the knife until the copper cladding has been cut through.

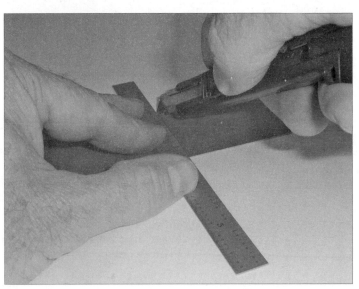

If the PC board has copper cladding on both sides (double-sided board) turn it over and score it on that side, directly under the first cut. Then break the board over the edge of your workspace or in a vice.

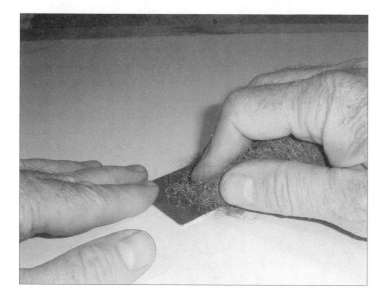

3. Trim any burrs or slivers from the edge of the board. Then clean one surface with steel wool or a synthetic scrubbing pad. The copper should be clean and bright.

4. Drill a ³⁄₃₂″ or 3mm hole in the PC board to mount the 7805 regulator chip in the center of the board. (If you have any thermal compound, put a film of it on the underside of the 7805 where the metal surface will contact the copper surface of the PC board.) Use a ³⁄₈″ to ½″ long 6-32 screw (6-32 is a standard screw size meaning the body size is #6 with 32 threads-per-inch), nut, and lock washer to attach the 7805 to the board.

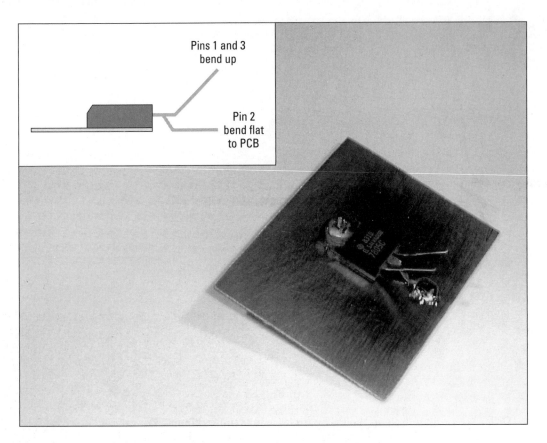

Pins 1 and 3
bend up

Pin 2
bend flat
to PCB

5. Bend the ground pin of the 7805 so that it is flat against the PC board surface and solder it to the PC board. Bend the two remaining leads up at a 45-degree angle.

6. Cut the cigarette lighter cord in half. Use your multimeter's resistance scale or continuity function to determine which of the wires goes to the ground contact of the lighter and to whichever contact of the output connector (if any) you wish to be grounded. Strip and tin 1/4" of each wire. If you are going to bring the cords through the walls of the enclosure, drill the necessary holes and insert the wires now.

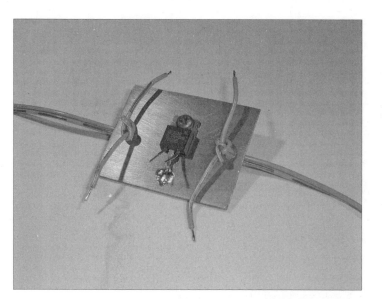

7. Drill holes at two edges of board big enough that the cord's cut ends will pass through. Insert the cords, input on the left, through the holes and tie a Western Union knot in the wires (see Chapter 9) to keep them from unraveling.

8. Solder the ground wires of the cords to the PC board.

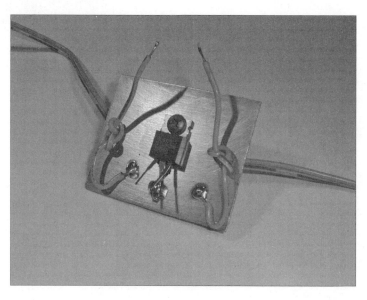

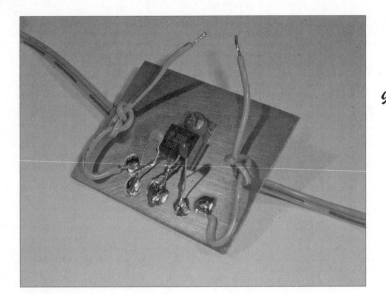

9. Tin the remaining leads of the 7805 and solder the capacitors from the input and output leads to ground. C1 connects from the input lead to the PC board and C2 connects from the output lead to the PC board. Be sure to connect C2 with the proper polarity.

10. Trim the leads of D1 to about 1/2″ long and tin them. Solder the cathode lead (the package has a stripe next to the cathode lead) to the input pin of the 7805 with the diode parallel to the PC board surface.

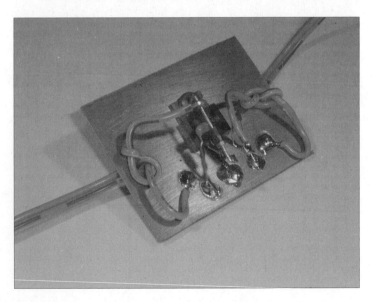

12. Test the regulator by inserting the cigarette lighter plug into the car's socket. (You may have to turn on the vehicle's ignition for voltage to be present at the cigarette lighter.) You should measure a steady 5V at the regulator's output between the positive output lead and ground (the surface of the PC board.)

11. Solder the positive input wire to the anode lead of D1. Solder the positive output wire to the output lead of the 7805.

Don't connect this regulator directly to your car battery without adding in-line fuses as shown in Chapter 7! A short circuit in the regulator will cause the input cord to overheat and possibly start a fire.

If there is no output voltage at the output pin of the regulator (measure with the voltmeter's negative lead connected to the PC board surface), confirm that there is input voltage from the cigarette lighter plug and that D1 is installed with the proper orientation. If the output voltage is 0 or significantly lower than 5 V and the regulator gets hot, you may have a short circuit in the output or C2 has been installed with the wrong polarity.

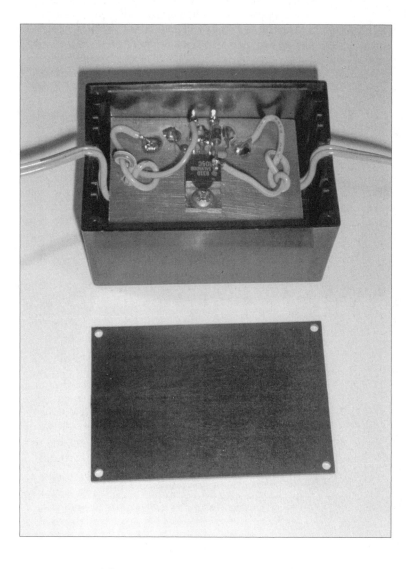

13. Put the assembled regulator into the enclosure. In the photo, a utility knife was used to make notches in the enclosure's edges slightly smaller than the input and output cords. When the enclosure's top was attached it will capture the cord between the top and sides, providing some strain relief for the cords. If the hole winds up big enough for the wires to move freely, add some hot glue on the inside of the enclosure to keep them from moving.

14. Label the regulator and document your work in your notebook.

Constructing the Alarm

A portable or temporary alarm combined with some kind of sensor can be a handy gadget. The alarm shown in this task can be used with a thermostat, a tip-over detector, a magnetic intrusion sensor — any type of sensor that acts like a switch and makes a connection between two terminals when it is tripped or activated. This task combines a speaker and an audio oscillator into a simple alarm system with the sensor of your choice.

1. Start by assembling the oscillator kit as directed by the manufacturer's instructions. (Chapter 4 contains additional information on building circuits on PC boards.)

TIP

When assembling kits, add your own sockets if they're not provided. This makes replacing a chip easy.

2. Collect the parts for the project and determine how your oscillator kit can be mounted in the enclosure. The kit used here does not have mounting holes, so a stick-on wire clip was used. These can be found in the electrical section of hardware stores.

3. Mark, centerpunch, and drill pilot holes for the speaker mounting screws and the phone jack. Use a hand reamer (see the task "Constructing the Level Controller" earlier in this chapter for information on using a reamer) to enlarge the hole for the phone jack. Mount the speaker and phone jack in the enclosure.

Hardware and appliance stores are great sources of inexpensive and novel mechanical components for experimenters. Show your project to a clerk and explain the problem you're trying to solve; he or she may have just the right part!

Be sure to use lock washers when installing any kind of vibrating component. Long periods of use can cause nuts to unscrew themselves from the vibration.

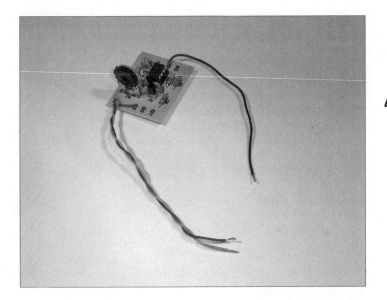

4. Cut some pieces of hookup wire and solder them to the assembled kit for power, ground, and audio output connections. The required length of the wires depends on the size of your enclosure and on how the various pieces are mounted in it.

5. Mount the assembled kit in the enclosure, using whichever method you've selected. The figure shows the kit held securely in an adhesive-backed plastic clip attached to the lid of the enclosure.

6. Solder the wire that supplies the kit with positive voltage to the phone jack's tip terminal. Solder the ground wire to the phone jack's sleeve terminal. (See Chapter 10 for more information on phone plugs.) Solder the kit's output wire to one of the speaker terminals — either will do.

7. Solder the 10μF capacitor's positive lead to the remaining speaker terminal and the capacitor's other lead to the phone jack's sleeve terminal (ground).

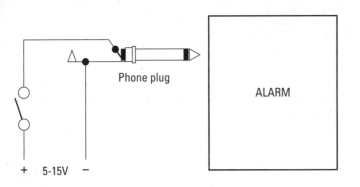

Phone plug

ALARM

+ 5-15V −

Close switch to
sound alarm

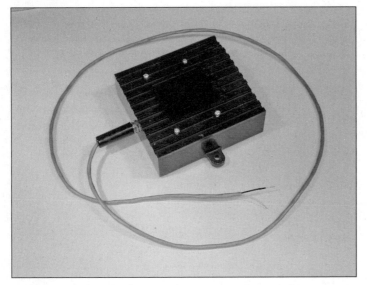

8. Connect +5 to +15 volts to the assembled alarm by connecting a two-conductor cable to the phone plug. Whenever voltage is applied to the alarm (such as through an alarm switch, as shown in the figure), the speaker will sound. The kit used in the figure has an adjustable tone, as well.

Chapter 6

Building from a Published Schematic

You're reading the latest electronics magazine and there's an article exactly right for your latest project. The author describes a circuit that will do just what you need to finish the job! Schematics are a common feature of many articles, both on the Web and in magazines. But those are just articles — there's no kit available, not even a printed-circuit board! You can build that circuit anyway by using one of the three techniques in this chapter: dead-bug style, Manhattan style, or wire-wrap. These methods have been used for years, even decades! Ready to give it a try? Let's go!

Even if a printed-circuit board (PC board) is not available for the circuit, the author may offer a kit of parts, a programmed microprocessor, or hard-to-find special parts. Print magazines have also begun adding supplemental material on their companion Web sites.

What types of circuits are suitable for building with the techniques in this chapter? As a beginner, you should start out simple, limiting yourself to building circuits with a few ICs at most. If the circuit has digital signals, they shouldn't be faster than a few megahertz. For analog or linear circuits, avoid those that are complex, very sensitive, or for which layout is critical. Work your way up to the fancier and more difficult circuits. There is a built-in limit to this process as more complex circuitry often requires a PC board, meaning that there should be one available for you to order.

Don't forget to keep your shop notebook handy while you're working! You never know when you'll have a good idea or need to record some measurements. When you're done, be sure to document what you built for later reference. "Tomorrow" is never a good time to write things down!

What method should you choose from the three presented here? None of the three is suited for high-power designs — those that involve currents of more than an amp or two, particularly wire-wrap which can only accommodate currents of around 100 mA at most. For high-current and high-voltage circuits, stick with point-to-point techniques that use separated terminals on insulating strips.

If you want to build a simple circuit that has one or two ICs, use the dead-bug technique. If the design has a lot of discrete components, such as a transistor circuit might have, then Manhattan style is more appropriate. Complex circuits and especially digital circuits are best done with wire-wrap, but don't try high-speed digital circuits or sensitive analog circuits with critical layout or noise constraints. If you really want to compare the techniques, build one circuit using each of the methods!

Caveat builder

Caveat emptor (buyer beware) is the rule on schematics in magazines and especially on the Internet. It's unlikely that the design has been subjected to rigorous testing or even had the benefit of a good initial production run, such as a commercial kit might receive. Publishers have also been known to make erors! Here's how to protect yourself against some of the more common problems:

Start by looking carefully for corrections on the same Web site or on the magazine publisher's Web site. Remember that it takes at least one issue and sometimes more than one for reader feedback to be published in a print magazine! In the meantime, use Internet search engines to look for comments on the circuit.

Trace out the schematic and compare it against data sheets for all the significant parts and ICs. This will catch mis-assigned pins and maybe turn up some questionable uses. Make sure you understand the purpose of every component on the schematic. (This is a great way to learn more about electronics!) If in doubt, write the editor or author. They'll appreciate hearing from you, an interested reader, if you're respectful and present your questions clearly.

All the checking still can't fix a poor design and some do get into print. For example, parts may be running too hot, undetected switching transients can cause premature component failure, a high-gain amplifier may want to oscillate, and so forth. Just be aware that the design may be flawed and be willing to test what you build.

Beware of obsolete and critical parts. The author may have had some in the junk box, but you may not be able to find them anywhere! (If you have a pressing need for obsolete parts, turn to Rochester Electronics at www.rocelec.com. Their company motto is "Leaders on the Trailing Edge of Technology" and they stock millions of discontinued parts.) Some parts may have unique characteristics without which the circuit won't work or will work poorly. A part may have a critical value as well or the circuit may require special adjustment and calibration procedures.

None of these should frighten you away from building circuits you find in magazines and on Web sites. Start simple — the schematics of simple projects are almost always correct. As you build up your experience, you'll feel comfortable with more and more complex designs.

Preparing to Build

Once you've researched your circuit and developed a complete parts list, you're ready to . . . get ready. Unless you have a very well-stocked shop, you're going to have to gather the necessary resources.

1. Develop a complete parts list, including all sockets and hardware. Make sure you have enough supplies of consumable materials such as terminals and wire before you start.

2. Order all required parts and obtain all necessary tools to build the circuit.

3. Determine the size of the PC board or wire-wrap board you'll be using.

4. Lay out the circuit to scale on paper first. It recommended that you use a scale of 2:1 or 4:1 so the drawing is easy to make and read.

5. Cut the board on which you'll build the circuit. Drill any necessary mounting holes before you begin constructing the circuit.

6. Make a plan of all the major steps you'll need to take, including testing.

7. Clean off your workspace and make sure you have all the tools and supplies that will be needed.

8. When the parts arrive, you'll be ready to go!

Building a Circuit Dead-Bug Style

Okay, okay, I'll cut right to the obvious question. It's called dead-bug style because the ICs are all mounted upside down on the PC board with most of their legs sticking up in the air! Figure 6-1 shows why — they look like a lot of . . . dead bugs! Another name for this type of construction is "air circuits" because the components are attached directly to each other over the underlying PC board. The technique is easy for beginners because it requires no special tools, gadgets, or techniques. Just solder the parts together and to a PC board and go! This is a great method to wire up a simple circuit to see how it really performs.

When building ala dead bug, the PC board surface is usually made to be the circuit's ground. This is sometimes referred to as a *ground plane*. Because the ground is literally everywhere and it's such a wide conductor, dead-bug construction works well for sensitive circuits and even at high frequencies. Conversely, this makes routing power supply connections a little more difficult, since everything is exposed.

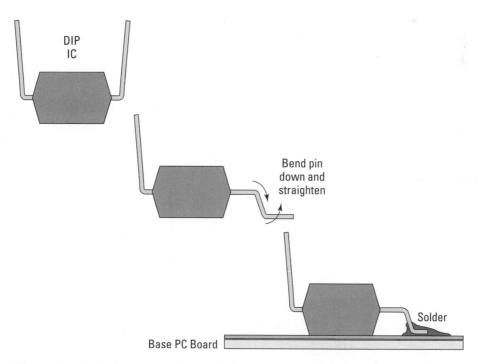

Figure 6-1: In dead-bug construction, ICs are turned on their backs with the pins extending up. Some pins may be soldered directly to the PC board surface. Most are soldered to components. Other components may also be soldered to the PC board, usually wired to be at ground voltage.

Dead-bug construction generally becomes impractical once the IC pin count exceeds 20 pins or if there are more than two or three ICs. Complex circuits have so many connections and components that it becomes hard to keep all of the connections straight and separated. Dead-bug style works best with one or two ICs.

Many of the components will attach directly to a pin of the IC (see the schematic in Figure 6-2). For this reason, the DIP ICs are greatly preferred in dead-bug construction. Surface-mount and leadless ICs (PLCC, QPCC, ball-grid array packages) are generally unsuitable for dead-bug construction.

If your circuit doesn't have a lot of components that attach to ground, you can use high-value resistors (1 MΩ or higher) as supports for other components. Solder one lead of the resistor to the PC board and stand it upright. Other components can then use the remaining lead as a *tie point* for connections with other components. Terminal strips (metal solder lugs mounted on an insulating strip — www.abbatron.com/products/?dir=/part/get/hdw_tsb_solb) can also be used.

To make the connections, components are mostly soldered directly together, lead to lead. You will have to bend the leads and orient the components to make the circuit work electrically and for mechanical stability. If jumper wires between connections are needed, bare wire can be used for short distances, but insulated #30 wire-wrap wire (see the final task "Constructing the DC-to-DC Converter Circuit" in this chapter) or #24 or #26 hookup wire can be used. Solid wire is preferred because it will keep its shape, whereas stranded wire is more flexible.

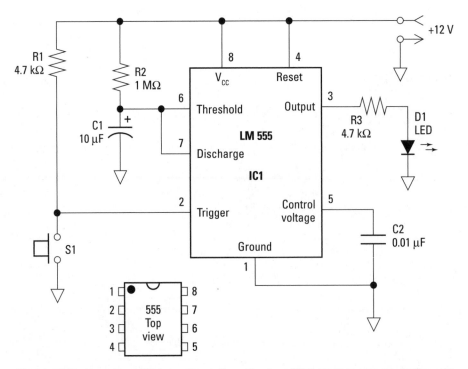

Figure 6-2: The schematic shows the configuration for a 555 timer IC to output a single positive pulse whenever switch S1 is closed. The length of the output pulse is 1.1 × R2 × C1 seconds.

The drawbacks of dead-bug construction

Dead-bug construction is very straightforward and requires no special tools, but it's not suitable for all kinds of circuits and application. In the spirit of helping you choose the appropriate technique for your circuit, the primary weaknesses of dead bugs are listed here:

✔ **Not mechanically robust.** Dead-bug circuits, like their namesake are easy to squash! They are also not suitable for tough environments like in cars or outside. The technique is a good way to build a prototype or experiment.

✔ **Hard to make major modifications.** While it's easy to change a component like a resistor or capacitor, it's quite difficult to change an IC or redesign a large chunk of the circuitry. It can also be difficult to restore all of the connections as the changes are made.

✔ **Solder blobs.** Dead-bug technique puts a premium on soldering technique. Messy soldering that drips excess solder will create short circuits, usually in places difficult to access. Use the minimum amount of solder and take your time.

✔ **Hard to duplicate.** When you're done, you will probably have a one-of-a-kind circuit. If that's all you need, great! Don't plan on using this technique on a production line, though.

When complete, a dead-bug circuit can be a real piece of sculpture. Or it can be a mess. It's up to you how it looks. Electrically, arty and messy will probably perform about the same, so it's a matter of your personal preferences how much time you spend arranging the components, whether wires are bent at right angles, and that kind of thing. My personal preference is for a neat appearance, but it takes longer.

Building Circuits Manhattan-Style

What? Another obvious question? Well, okay, this technique is called Manhattan style because small squares of PC board material are glued onto the larger piece of PC board. The small squares act as insulated tie points for soldering circuit components together. The result looks a bit like a city map with the squares representing city blocks and the space between them the streets.

Where do the pads come from? You make them, of course! It's easy! Start by cutting a strip of PC board material about ¼" wide. (The material can be single- or double-sided.) Use the same technique for scoring and breaking the PC board as described in Chapter 5's task "Building a 12V-to-5V Regulator." Take a pair of medium-sized wire cutters, heavy scissors, or tin snips and cut the strip into small, square pieces as shown in Figure 6-3. Once you get started, it's a good idea to make a large number of pads for future use.

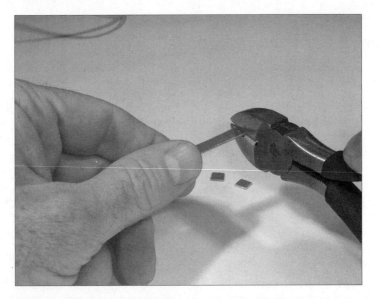

Figure 6-3: Cutting a strip of PC board material into small squares creates isolated pads. The pads are glued onto a base of PC board to be used as tie points for circuit connections. The "block and street" grid pattern of the pads on the base material gives rise to the name Manhattan-style construction.

While this task will use square pads — and most circuit builders do use square pads — there is no reason that they have to be square. You can use a heavy-duty hole punch to cut round pads, the pads can be rectangular, or even oddly shaped. In fact, long narrow strips make good power supply rails, just as for the solderless breadboard in Chapter 3! Take advantage of some spare time to make a couple dozen pads of different sizes and shapes, storing them in small zippered plastic bags.

Pads are usually attached in rough alignment with the circuit schematic shown below in the task "Constructing a Shortwave Buffer Circuit". It is not necessary to attach all of the pads at once, since as the circuit is built, it's often necessary to adjust the layout for convenience. When you prepare the schematic for building, go through it and outline the connection points that will use a pad to help you visualize pad placement.

Manhattan-style construction works best when the circuit is composed primarily of discrete components, such as resistors, capacitors, transistors, diodes, and so forth. If ICs are present, they aren't connected directly to the pads, but other components may be. This creates a hybrid style of Manhattan style and dead-bug style! This type of construction can be used at HF (High Frequency) with good results, even though each pad adds a little capacitance to ground.

The pad-based technique rapidly becomes untenable when you have more than 25 pads or more than a couple of ICs. Remember that this technique is not meant for mass production or for building miniaturized or complex equipment. It's a useful way to build one of something or give a new design a try when you expect to be doing some experimenting.

The drawbacks of Manhattan-style construction

Manhattan-style construction works really well for small circuits, particularly those made with individual components. It definitely runs out of steam for complicated and high-frequency circuits, so you should be aware of its limits.

✔ **Pads take up room.** They do, but miniaturization is not usually a goal of Manhattan-style construction. Use a large enough piece of PC board as a base so that building is convenient, thus avoiding mistakes that result from building in a cramped space.

✔ **Pads add capacitance to ground.** Each pad acts like a small capacitor to ground (the PC board surface) and this can be a problem for proper circuit operation. Since Manhattan-style construction is generally not used for circuits operating at VHF (above 30 MHz), this usually isn't a problem but you should be aware of it.

✔ **Glue can get messy.** This is not a project on which to learn how to use a hot glue gun! Use a miniature gun as a big one will leave drips and threads of glue all over the board and pads.

✔ **Labor intensive to make pads.** Pads are not sold in stores, so what choice do you have? This is a perfect activity while listening to that new CD you just bought or while stuck inside on a rainy day.

Building Circuits Using Twist 'n' Twirl Wire-Wrap

No mysteries exist about the reason for the name of this circuitbuilding technique! Wire-wrapping was developed in the 1960s for building telephone system equipment. It rapidly became popular for building complex circuits that needed to be modified (such as for prototypes) or that were only constructed in low volumes. Wire-wrap can even be automated, building large circuits composed of dozens of ICs on a single board! (Additional information about wire-wrapping is available in Wikipedia at http://en.wikipedia.org/wiki/Wire_wrap.)

A connection made by wire-wrapping is shown in Figure 6-4. Solid copper wire, plated with silver or some other soft, non-oxidizing metal, is tightly wrapped around a square metal pin. The corners of the pin bite into the wire very tightly, keeping the joint tight. Without some kind of strain relief, the sharp bend around the pin's edge would cause the solid wire to eventually crack. To cushion the wire, one or two turns of the insulated portion of the wire are first wrapped around the pin at the bottom of the wrap.

Because the connections are made with insulated wire, the ICs in the circuit can be very close together. The wires connecting their pins are routed along the rows between them. This allows for very densely packed circuits, just right for digital logic and microprocessors.

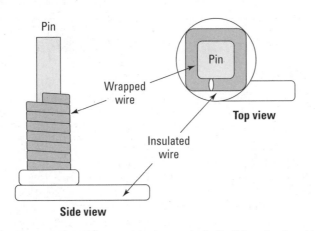

Pin

Wrapped
wire

Pin

Insulated
wire

Top view

Side view

Figure 6-4: Wire-wrapping is a method of building circuits without soldering. A solid copper wire is wrapped tightly around a square pin whose edges bite into the wire for a secure connection. A turn or two of insulated wire at the bottom of the wrap provides strain relief for the wire.

Wire-wrapping requires special tools and wire as shown in Figures 6-5 and 6-6. Besides some kind of wire-stripping tool or a combination stripping/wrapping/ unwrapping tool, you'll also need wire-wrap sockets with long pins that extend through an insulating board. You'll mount discrete components such as resistors and capacitors to a *socket header* or wire-wrap pins before plugging it into a socket. A must-have material is *perfboard* made of an insulating plastic or resin material and drilled with a grid of holes 0.1″ apart. Figure 6-5 shows wire-wrap labels that slip over the socket's pins, holding the socket to the board and labeling each pin by number. Next to the pin label is a pair of pins that can be pressed into the perfboard and used to connect to individual wires or components. Most of the wire-wrapping you will do uses these tools and materials.

Figure 6-6 shows a motorized wrapping tool. Most circuits can be made using the manual combination tool shown in Figure 6-5. If you do a lot of wire-wrapping, however, you'll definitely want to purchase the motorized wrapper. It makes a complete wrap in less than a second!

The wire used for wire-wrapping is not ordinary solid hookup wire. The wire itself is a specific size: #30. Most stripping and wrapping tools are designed to work with this standard size. The wire is also coated with silver or some other soft metal that will make a good connection with the sharp edges of the wire-wrapping pins. If you use bare copper wire, for example, the connections will eventually degrade and the circuit won't work properly. The insulation, called *Kynar®*, is formulated to break cleanly and evenly when stripped from the wire. Use wire-wrapping wire to avoid a lot of unnecessary headaches.

Pin labels come in various sizes to fit all of the common socket sizes; from 6-pin to 64-pin. Using a pin label to hold the socket against the board until a wire or two has been attached allows you to avoid using glue. While you can use a cyanoacrylate "super" glue, you don't want to get any glue on the pins as you slide the socket onto the board; this can cause bad connections that are hard to find. The label will hold the socket firmly enough to keep sockets seated firmly against the back of the perfboard.

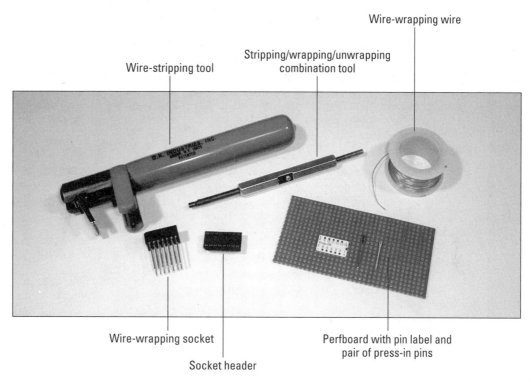

Wire-wrapping wire

Stripping/wrapping/unwrapping
combination tool

Wire-stripping tool

Wire-wrapping socket

Socket header

Perfboard with pin label and
pair of press-in pins

Figure 6-5: Wire-wrapping tools and materials.

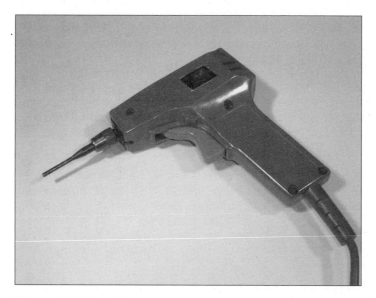

Figure 6-6: A motorized wrapping tool saves a great deal of effort when constructing complex circuits or if you will do a lot of wire-wrapping.

The drawbacks of wire-wrapping

For complicated digital circuits there's no better prototype technique than wire-wrapping; that's why it's been around for nearly 50 years! Before you decide to build everything using wire-wrap, you have to know where it *doesn't* work well.

- ✔ **Making changes requires changing the wire.** If you decide to change a connection, you should change the entire wire for the best reliability of the connection. A reused wire-wrap wire does not wrap around the pin as tightly. In practice, most casual builders try to use the wire by straightening it and rewrapping. A second re-use is almost impossible, though.

- ✔ **Takes some practice.** There is "kindy a knack to it" as my great-aunt Lexie used to say. You will rapidly learn to strip the wire, get it in the right hole of the wrapping tool, and make each wrap evenly. Once learned, you'll be surprised at how quickly you can create even big circuits!

- ✔ **Extra expense of sockets and materials.** The convenience and flexibility of wire-wrap comes at a small price. You'll have to keep the sockets, labels, wire, and pins on hand along with the necessary set of tools.

- ✔ **Not suited for sensitive circuits.** The wires making the connections of a wire-wrapped circuit lay directly on top of each other, making separation of signal paths more difficult than for other types of construction. This can cause feedback and *cross-talk* (contamination of one signal by another) in sensitive circuits.

Later in this chapter, you will build a simple DC-to-DC converter circuit. The star of this circuit (shown in Figure 6-7) is the ICL7662 IC. It contains the necessary switches and clock signals to perform the electronic sleight-of-hand that creates the negative voltage. This particular circuit uses four of those switches. The chip first connects C1 to the +12V input and ground, charging it up to +12V. Those connections are opened, leaving the capacitor full of charge, but not connected to anything. The positive terminal of C1 is then connected to ground and its negative terminal to pin 5. Some of the charge stored in C1 is then transferred to C2, but with the opposite polarity from the input power. Then connections are opened and the process is repeated, just as if a bucket (C1) was being used to carry charge from a tap (the input power) to a tank (the output capacitor, C2).

The ICL7662 does this so quickly (each step takes only about 10 μs!) that to the circuit being supplied with negative voltage, it looks like a regular power supply is connected. The only limit on output current that can be generated is from the size of the capacitors and the speed at which the chip works. The ICL7662 can do several other neat tricks, too. Download its data sheet from www.datasheetcatalog.com/datasheets_pdf/I/C/L/7/ICL7662.shtml to learn about this interesting device.

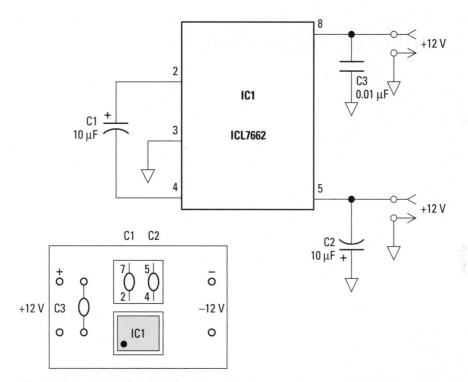

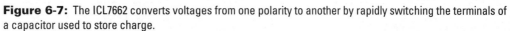

Figure 6-7: The ICL7662 converts voltages from one polarity to another by rapidly switching the terminals of a capacitor used to store charge.

Constructing the Timer Circuit

Stuff You Need to Know

Toolbox:

✔ Fine needle-nose pliers, wire cutters, soldering iron and solder, straight-edge, utility knife, small vise

Materials:

✔ Single- or double-sided PC board
✔ NE555 timer IC
✔ Two 4.7 kΩ and one 1 MΩ, ¼-watt resistors
✔ 10 µF, 16V electrolytic capacitor
✔ 0.01 µF, 25V or higher ceramic capacitor
✔ LED (any color)
✔ Pushbutton or other switch that closes only while pressed (called *momentary*)
✔ 6V to 12V power supply or battery pack

Time Needed:

Less than half a day

This circuit uses one of the most popular ICs of all time, the 555 timer. There is a good description of how this circuit works at www.uoguelph.ca/~antoon/gadgets/555/555.html or at en.wikipedia.org/wiki/NE555. This simple IC can be wired up to output single pulses, a continuous stream of pulses, act as a time delay, and perform many other functions its designers never imagined. Still going strong after more than 30 years of life, the 555 timer should be a part of every electronic-er's bag of tricks!

In this example circuit, the timer is wired to output a single pulse whenever the reset switch is closed. This is the *monostable* configuration. The length of the pulse is determined by the values of C1 and R2. The output of the timer (pin 3) will be ON — at the power supply voltage (V_{CC}) — as long as the voltage on C1 is less than 2/3 V_{CC}. Assuming the switch is closed just long enough to discharge C1, the output will be $1.1 \times R2 \times C1$ seconds long. The LED will light whenever pin 3 is ON. With R2 = 1 MΩ and C = 10 µF, the duration of the pulse should be about 11 seconds.

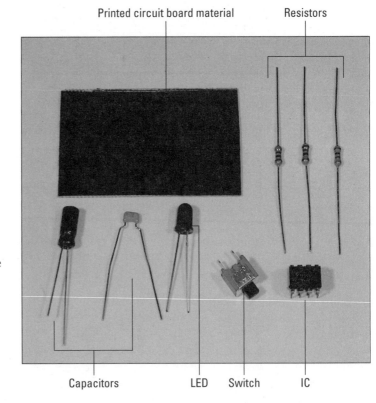

Printed circuit board material   Resistors

Capacitors   LED   Switch   IC

1. Start by making a photocopy of the schematic and making sure you have all the parts ready to go.

Use a fine (small-tipped) pair of needle-nose pliers. The larger needle-nose pliers are more suited to mechanical chores and are too coarse for the job at hand.

2. Cut and clean a rectangle of PC board material about 1½" square, as shown in the task "Constructing the Regulator" in Chapter 5. Size is not critical; it's better to have the board be a little too large than too small. You can use either single-sided or double-sided PC board. Cut the board by scoring it with a metal straightedge and sharp utility knife. Score the board on both sides and then bend it over the edge of your workspace or in the small vise.

3. Use the pliers to bend Pin 1 of the IC, as shown in Figure 6-1 earlier in this chapter. Solder Pin 1 to the PC board (ground). Mark off each connection on your schematic as you create the connection on the board.

4. Solder C2 to Pin 5 and to the PC board. Having two pins soldered holds the IC steady.

When soldering IC pins to the PC board, it works best to make a small bump of solder on the board, tin the IC lead, then place the IC lead on the solder and heat both together.

When soldering IC pins to the PC board, it works best to make a small bump of solder on the board, tin the IC lead, then place the IC lead on the solder and heat both together.

Remember that all the IC pins are mirror-imaged when the board is turned over on its back! It may help to make a pencil mark next to Pin 1 on the IC's bottom surface to help you orient yourself as you work.

When attaching an IC to the PC board, attach two or three components that will hold the IC steady. This keeps the IC from twisting, which can cause pins to break off or short out to nearby wires.

5. Bend Pins 4 and 8 of the IC over the body of the IC, and then solder a short piece of bare wire to them, leaving some wire sticking out for contact later. This will be the +12V connection to the circuit.

6. Solder one pin of the switch to the PC board, close enough to the IC so Pin 2 can be bent out to touch the remaining pin of the switch. Solder Pin 2 of the IC to the switch pin.

7. Squeeze Pins 6 and 7 of the IC together and solder them together. Then solder C1 from Pins 6 and 7 to the PC board.

Keep the bits of bare wire left over from trimmed components. They will come in handy for making short jumper connections.

If getting a good connection between wires or IC pins is difficult, just tack the component in place with a couple of light connections, and then go back and do a more thorough job of soldering when the component is held steady.

Leave plenty of room near the IC for connections, since this area gets crowded quickly. Tin the leads of components first for quicker, more solid soldering and sturdier mechanical connections.

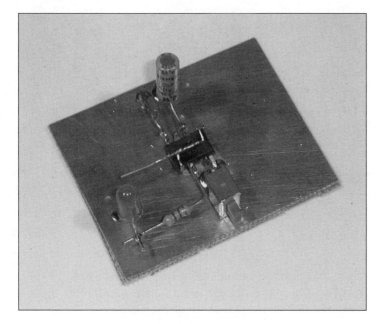

8. Solder R3 and the anode (for help identifying the LED leads see Step 5 of "Constructing the Digital Timer" in Chapter 3) of LED D1 together, then solder the other lead of R3 to pin 3 of the IC and the cathode lead of the LED to ground. (You can be sure you have the anode and cathode correctly identified by using your multimeter to test the LED in the same way you would test a diode.)

9. Solder R2 between C1 and the +12V wire connected to IC pins 4 and 8. Solder R1 from S1 to the +12V wire.

The junction of two wires should not be under tension when soldered. A junction that wants to pull apart may spring apart later when another wire is soldered on.

It's a little easier to make contacts with wires at right angles if the layout permits. Avoid *butt* contacts with the wire ends coming directly together, as they are very difficult to solder correctly. Overlapped end-to-end connections work fine.

10. Attach the wires that you intend to connect to the power supply.

11. Give your circuit a very close inspection, looking for balls of solder, loose wire bits, accidental short circuits, and unsoldered open circuits. Test the circuit by applying power. The first time power is applied, the LED will light until C1 charges to a level that equals ⅔ of the voltage at V_{CC}. When the LED goes out, the circuit is ready to work. Press the switch; the LED should light for about 11 seconds, and then go out again.

To help manage the rat's nest of wires and leads, twist any pairs of wires that carry power or ground as shown in Step 10's figure.

If your LED doesn't come on and your IC gets hot, you probably wired it backwards. This is a very common error for folks who are just learning the dead-bug technique. Unfortunately, the IC probably didn't survive either, so you'll have to get a new chip *and* rewire the circuit! If the IC isn't hot, but the LED still doesn't come on, watch Pin 3 with your voltmeter as you press the switch. If it doesn't show a positive voltage, check Pins 6 and 7 to see whether C1 is charging and discharging. Use the circuit's operation descriptions to guide your testing efforts.

Constructing the Shortwave Buffer Circuit

The shortwave signal buffer circuit shown in the following figure will be used to demonstrate Manhattan-style circuit construction. This simple transistor amplifier is used between a short receiving antenna, such as a telescoping whip, and a radio receiver's input. If you've ever used an audio transformer to connect a high-impedance microphone to a low-impedance PA system input, this circuit accomplishes much the same thing. It will *buffer* (meaning to isolate) signals across the lower-frequency shortwave bands up to a few megahertz. The output is suited for connection to the customary low-impedance, 50Ω radio inputs.

The circuit has no voltage gain at all, but does increase the output power because it can deliver the same voltage to a much heavier load. (You can find a good overall description of transistor functions and basic amplifiers at www.faqs.org/docs/electric/Semi/SEMI_4.html.) There may be somewhat better transistors to use than the 2N3904, but it is widely available and tolerant of the occasional abuse suffered during experimentation. As this is a demonstration project, the 2N3904 will do the job.

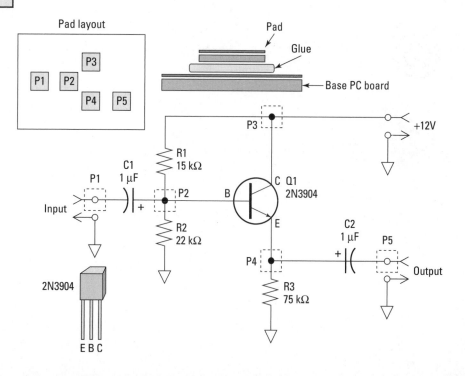

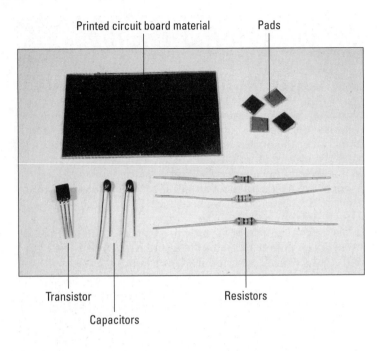

Printed circuit board material   Pads

Transistor

Capacitors

Resistors

1. Start by making a photocopy of the schematic and making sure you have all the parts ready to go.

2. Cut and clean a rectangle of PC board material about 1½″ square, as shown in the task "Constructing the Regulator" in Chapter 5. Size is not critical; it's better to have the board be a little too large than too small. You can use either single-sided or double-sided PC board. Cut the board by scoring it with a metal straightedge and sharp utility knife. Score the board on both sides and then bend it over the edge of your workspace or in the small vise.

TIP

Use a fine (small-tipped) pair of needle-nose pliers. The larger needle-nose pliers are more suited to mechanical chores and are too coarse for the job at hand.

3. Put a *small* dot of hot glue in the middle of the PC board, quickly attaching one of the pads at the position marked P2 (see Figure 6-4 earlier in this chapter). Repeat for pads P3 and P4 to form a triangle with about ½″ between pad centers. The 2N3904 transistor leads will be soldered to these pads. Remove any strings or drips of glue as you go.

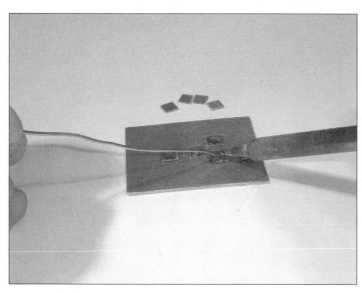

4. Hot-glue the input and output pads, P1 and P5, to the base PC board.

5. Tin all pads. Keep the iron on the pad just long enough to melt some solder and make a small pool. Use your multimeter to check for and remove accidental short circuits between the top of the pad and the base PC board.

6. Tin the leads of the transistor, spread them apart slightly, and flare the ends of the leads so the last ⅛″ or so is parallel to the surface of the PCB. Solder them to pads P2, P3, and P4 as shown in the photo.

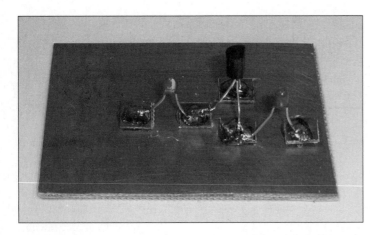

7. Solder C1 between pads P1 and P2 and C2 between pads P4 and P5. Be careful to orient them in the correct polarity.

TIP

Have the pad ready in one hand while putting the drop of melted glue on the PC board and place it quickly. The glue will solidify rapidly as the copper coating of the PC board conducts the heat away. Don't worry about getting the pad square with the other pads — that doesn't matter just now, and you'll get a chance to realign the pads slightly in the next step.

8. Solder R1 between pads P3 and P2 and R2 from pad P2 directly to the PC board as ground.

9. Solder R3 from pad P4 directly to the PC board as ground.

The tinning will melt the glue, so don't move the iron around, just press the pad against the board. Use the tip of the iron to align or rotate the pads, if necessary. Don't move them around very much as that will just smear the glue. It doesn't matter if the pads don't all wind up square and neat — the electrons won't care!

10. Attach the power leads to pad P3 and directly to the PC board as ground. Connect the input and output signals (shown in the figure as miniature coaxial cables) at pads P1 and P5, respectively.

11. Test the circuit by listening to your radio while turning the circuit's power ON and OFF. You should be able to hear a change in the background noise level, at least. If you can tune in a weaker signal, such as a distant AM broadcast station, the signal should be louder or clearer with the buffer circuit turned ON.

If your buffer makes signals weaker or less clear, you may have the transistor hooked up improperly or one of the capacitors hooked up backwards. Double-check your connections and make sure you have all the resistor values correct. There should be about 0.6–0.7V from the transistor's base to emitter leads. The emitter voltage to ground should be about 0.3V and the collector voltage to ground the same as your power-supply voltage.

Constructing the DC-to-DC Converter Circuit

In this task, you'll create a handy little circuit: a DC-to-DC converter that changes a positive voltage to a negative voltage. It's very common to need a small amount of power at a negative voltage, and adding another power supply can be quite expensive. By providing that power with a DC-to-DC converter, the job is done at a lot less expense. This simple circuit is a good way to learn the basics of wire-wrapping. Refer to Figure 6-8 for the schematic.

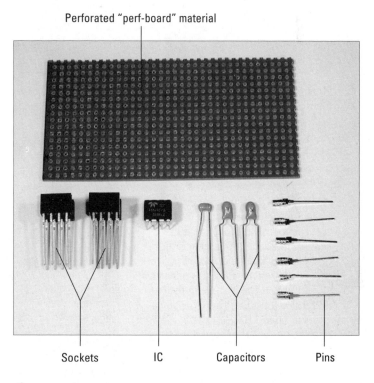

Perforated "perf-board" material

Sockets   IC   Capacitors   Pins

1. Start by making a photocopy of the schematic and making sure you have all the parts ready to go. Layout is not critical to this circuit, but a suggestion is shown in Figure 6-8. Placing the input and output connections at opposite sides of the perfboard and the two sockets side by side in the middle helps organize the connections.

2. Score and break a piece of perfboard about 1½" square, using the same technique as for the Chapter 5 task "Constructing the Regulator."

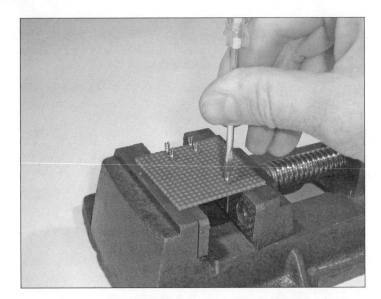

3. Press the separate pins into the board, using a small screwdriver.

4. Place one of the sockets on the board, noting which is pin 1. If you have a pin label, use it, otherwise place some tape across the sockets to hold them on the board.

5. Prepare the first wire by cutting a piece of wire-wrap wire 3" long. Use the stripping tool to strip ¾–1" of insulation from each end of the wire. This is done by pressing the wire into the narrow slot between the stripping tool's fingers and pulling the wire out. You may need to use needle-nose pliers to grasp the insulation; the wire is quite thin.

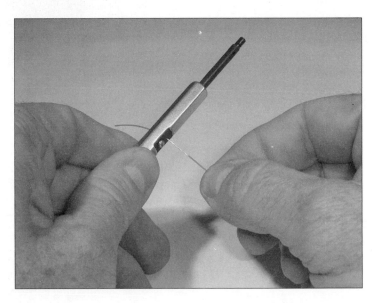

When placing sockets on the board, they should all have pin 1 oriented in the same direction. This makes it easier to keep pins identified. If you have some sockets at right angles to others, keep the orientation of the sockets consistent in each group.

As with dead-bug construction, the pin layout is mirror-imaged from the Top View.

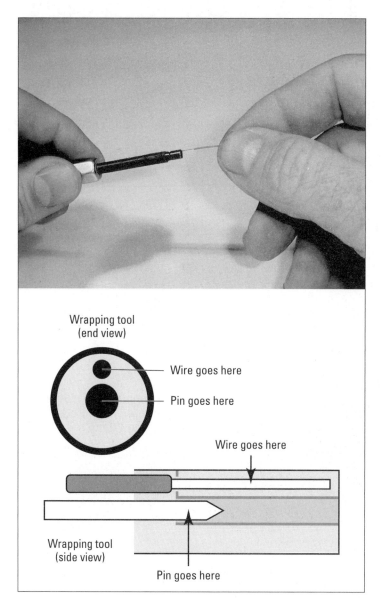

Wrapping tool
(end view)

Wire goes here

Pin goes here

Wire goes here

Wrapping tool
(side view)

Pin goes here

6. Insert one stripped end of the wire into the end of the wrapping tool, as shown in the figure. The drawing shows where the wire should go and where the pin will go.

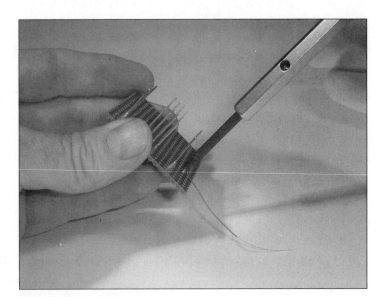

7. Slide the tool holding the wire down onto the positive-input pin. It should go all the way to the bottom.

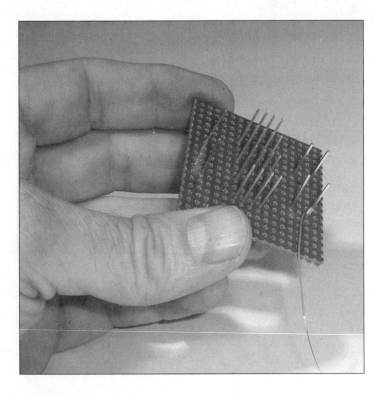

8. Turn the wrapping tool clockwise until the wire is completely wrapped around the pin.

The standard direction for wrapping is clockwise, just as for tightening a screw or nut.

9. Insert the other end of the wire into the wrapping tool and repeat for the adjacent pin of the bypass capacitor, C3. If you're unsure of your technique, this would be a great time to practice, wrapping several wires until you feel like you've got the hang of it. Then remove them (as described in the sidebar "What Happens If You Change Your Mind?"); the pins won't wear out and can be wrapped over and over again.

What happens if you change your mind?

One advantage of wire-wrapping is the ease of making changes or correcting wiring errors. To that end, the makers of wire-wrap tools have also created an *un*wrapping tool. If you're using the combination tool, the unwrapper is on the opposite end from the wrapper. In fact, you may have tried to insert the stripped wire into the unwrapper at least once (you're in good company). If you have a motorized wrapping tool, you'll need a separate unwrapping tool.

To remove a wrapped wire, slide the unwrapping tool onto the pin until it contacts the wrap. Then turn the unwrapping tool in the opposite direction from that used to wrap the wire onto the pin. The usual direction to unwrap is counter-clockwise. The combination tool just pushes the wire away from the pin as it unwraps. A separate unwrapping tool has a hook that actually unwinds the wire, making it a little easier to reuse the wire at its new destination.

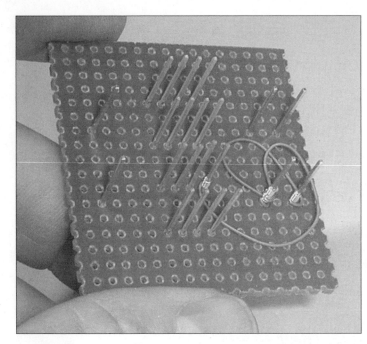

10. Repeat the process to connect the same pin of C3 to pin 8 of the IC socket.

 Wire-wrap wires are usually cut somewhat longer than the path between pins. This allows plenty of wire to form the initial cushioning wraps and it's usually more convenient than trying for a direct route. The extra length allows you to place the wires between (and away from) the pins as well.

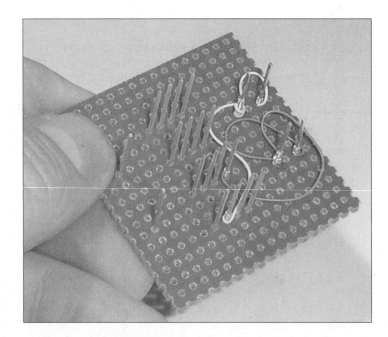

11. Use another pair of wires to connect the negative input pin to the remaining pin of C3 and to Pin 3 of the IC socket.

12. Connect Pin 5 of the IC socket to the negative output pin.

You can see why it's a good thing to have different colors of wire: If everything is just white or black or red, it can be very difficult to trace connections and keep everything straight.

13. Connect the leads for C1 by connecting pin 2 of the IC socket to pin 2 of the capacitor socket (C1 positive lead). Connect Pin 4 of the IC socket to pin 7 of the capacitor socket (C1 negative lead).

14. Connect the leads for C2 by connecting Pin 4 of the capacitor socket (C2 negative lead) to the negative output pin. Connect Pin 5 of the capacitor socket (C2 positive lead) to the output ground pin.

15. Finish the ground connections by connecting pin 5 of the capacitor socket (C2 positive lead) to the input ground pin.

16. Install the IC and capacitors in their sockets. Use needle-nose pliers to insert the capacitors' leads into the sockets.

If the circuit has a lot of discrete components, solder them to a DIP header and then insert the header into a wire-wrap socket.

17. Solder the bypass capacitor, input, and output connections.

18. Test your circuit by applying power to the input connections and measuring the output with your voltmeter. The output voltage will be slightly less than the input voltage, but with the opposite polarity.

If your voltage converter doesn't work properly, check to make sure the IC and capacitors are inserted correctly into their sockets. It's quite easy to get things reversed between the top and bottom of the board! Also be sure that capacitors C1 and C2 were installed with the correct polarity. If you still have no output, carefully trace your connections and use a magnifying glass to look for accidental short circuits between socket pins.

Part III
Cables and Connectors

The 5th Wave   By Rich Tennant

In this part . . .

Cables and connectors are often omitted from books about electronics, but they're virtually omnipresent! *Circuitbuilding Do-It-Yourself For Dummies* takes the opposite tack, giving these important elements of circuitbuilding their very own set of chapters. By learning how to work with cables and connectors, you can save a ton of money by making and repairing them yourself!

We begin by learning how to install many common terminals and crimp-on connectors. You'll be introduced to the inexpensive crimping tools that pro technicians use to install connectors quickly and securely. After you've mastered crimping, you can move up, up, up in frequency to the radio-frequency (RF) cables and connectors used for video and radio equipment.

Connecting electronics to sources of power has to be done properly to be safe and to provide high-quality power. By learning how to perform some common power wiring tasks, you cultivate the confidence that you've done the job correctly. We also show you examples of how sensitive audio and sensing circuits present their own special requirements that distinguish them from power connections. We address those requirements and guide you through building your own temperature sensor.

Chapter 7

Terminals and Connectors

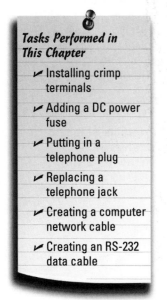

Most electronics books and how-to guides spend a lot of time inside the box, but how do you get the box hooked up? Taking a look behind any stack of electronic stuff and what do you see? Wires and cables! On each end of each wire and cable is a connector or some terminals. As a successful electronic-er, you'll need to know how to select, create, maintain, and repair those simple-but-crucial items that get the signals and the juice where they need to go. In this chapter, we take a look at several common wire-cable-connector combinations. Dozens more are possible, but if you learn these basic techniques, you'll be able to figure out others easily.

One consideration that applies to all such combinations: Use the right wire or cable. If you're working with digital data or audio signals, you'll need to use a data or audio cable if you want the data to flow fast and the audio to remain crisp and clear. If the cables and connectors are important to making a good quality connection, the equipment manufacturer will specify what to use; usually in an operating or user manual.

Crimp Terminals and Tools

As electrical appliances and machines became widespread, electricians and engineers tried a lot of different ways to make reliable connections in power and control circuits: screws, soldering, and clamps. They worked, but were often time-consuming. Finally, someone hit upon the idea of capturing a wire inside a tube by collapsing or *crimping* the tube. They found that if the crimps were made correctly and securely, the resulting connection was just as reliable as the traditional methods. Today, crimped-on terminals and connectors of all sorts are used in nearly every field of electronics. Each type of crimp is a partnership between a terminal or connector and a special tool that crimps the connection just so. Let's meet the three most common varieties.

Crimp terminals

Crimp terminals are the most common method of making connections for control and DC power wiring. You'll find crimp terminals in your car, in your home, and many appliances. Each terminal makes a connection for a single wire that is stripped and inserted into the terminal, then attached by crimping. Being able to install these terminals properly is important to prevent intermittent connections. Poor connections carrying high currents can also get hot enough to become a fire hazard. It's simple to install these terminals, however.

There are three common terminal styles seen in Figure 7-1; *ring*, *fork*, and *spade*. Ring and fork terminals are intended to be attached to screw terminals, usually on an array of terminals called a *barrier strip* or *terminal strip*. Ring and fork terminals are specified by *stud size* (the diameter of the screw terminal they fit — #6, #8, #10) and by wire size (the thickness of the wire that they can be crimped to hold).

There are a number of variations on each terminal style. Enter "crimp terminals" into an Internet search engine to find dozens of catalog links to terminal vendors with complete drawing and style information. Appendix A contains links to the Web sites of several distributors of electronic components, including terminals.

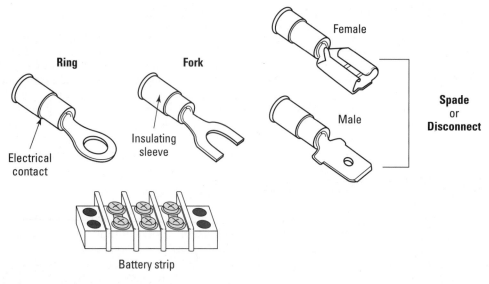

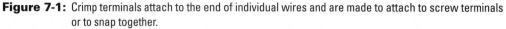

Figure 7-1: Crimp terminals attach to the end of individual wires and are made to attach to screw terminals or to snap together.

The various crimp terminals differ in the following ways:

✔ **Ring terminals** have a hole through a flat terminal. A screw is inserted through the hole so that the terminal can not be removed without also removing the screw or nut. This prevents a ring terminal from pulling out from under a screw due to vibration or tension.

✔ **Fork terminals** form a 'Y' that fits around a screw terminal and are held in place by the clamping force from the screw. They are used when vibration or tension are not expected to cause the terminal to loosen.

✔ **Spade terminals** come in male and female versions, also called quick-disconnects because they can be inserted and pulled apart without tools.

✔ **Butt-splice terminals** are used to permanently join two wires together end-to-end. They consist of a single tube that is crimped on each end to capture the two wires.

Terminals are available in insulated and non-insulated versions, with the insulation color signifying what size wire fits the terminal. Table 7-1 shows the three common terminal colors and sizes.

Table 7-1	Terminal Insulation Color and Wire Size	
Color	*Wire Size*	
Red	22–18	
Blue	16–14	
Yellow	12–10	

Using the terminal size required for the wire diameter being used is important to ensure a proper, low-resistance crimp in which the wire is completely captured by the terminal and will not work loose. Using terminals that are too large or too small will result in an unreliable connection.

Do not use crimp terminals with solid wire — and do not use pliers or other tools to crimp terminals. Solid wire is not captured firmly enough by a crimp; it will pull out or work loose. Solid wire may also eventually break at a nick or sharp terminal edge. Pliers and other such tools only deform the terminal body and don't apply enough concentrated force to properly capture the wire.

The terminal crimping tool

Industrial and commercial terminal installers use special tools that are made to make hundreds of crimps precisely the same. Those tools are too expensive for the hobbyist. By far the most common terminal crimping tool used by civilians is the hand-operated crimper-stripper shown in Figure 7-2. Although the crimper looks like a pair of pliers (and may have plier-type jaws at the tip), this tool can make a very satisfactory crimp if used properly: Be sure to use the correct position, orient the terminal properly, and squeeze until the crimp is well formed.

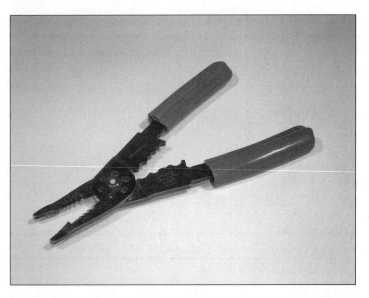

Figure 7-2: This inexpensive hand crimper will make acceptable crimps if used properly.

The keys to making a good crimp are threefold:

- ✔ Match the wire and terminal size.
- ✔ Strip the wire properly.
- ✔ Crimp the correct side of the terminal.

Figure 7-3 shows how the wire is inserted in the terminal and how the crimp captures the wire. Take note of the following in the figure:

- ✔ The tabs below the electrical contact area are rolled up to form the terminal's hollow barrel.
- ✔ The area of ring and fork terminals that makes electrical contact with the screw or nut stays flat.
- ✔ The cross-section of the barrel before and after a crimp shows how the wire is captured and solidly held.

In a proper crimp, the bottom non-seam side of the crimp tube is collapsed to form a short tab that presses the strands of wire against the inside of the rolled portion of the crimp tube. A properly crimped terminal captures all strands of the wire and the terminal wall tightly so as to resist corrosion and to provide high *pull-out* strength.

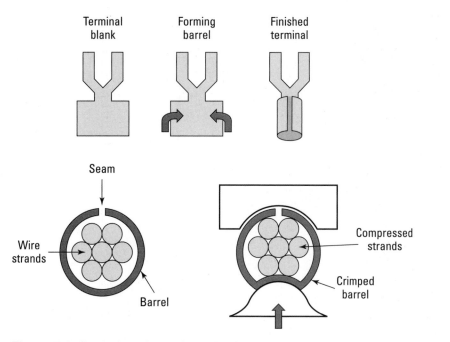

Figure 7-3: Terminals are formed from a flat sheet of metal.

If the seam side of the crimp tube is crimped, the walls of the tube flare apart, greatly reducing the terminal's gripping strength. Terminals crimped in this way will usually fail because the wire pulls out, or because the resulting connection is relatively poor and eventually fails from vibration, heating, or corrosion.

Take a close look at your crimping tool. Each stripping position is labeled with the wire size. Find the position for your wire and practice stripping just enough of the insulation off the wire. The bare wire should protrude no more than $\frac{1}{16}$" from the barrel of the terminal as shown in Figure 7-4. If your stripper doesn't strip the wire cleanly, move to the next smallest size and try again. Different types of insulation require different depths of cutting by the stripping edges.

Look closely at the crimping part of the tool. The crimping positions are located between the pivot and the handle. They are labeled by terminal size and may have colored dots to indicate the position for each size of terminal. Note which side of the tool the crimping tab is on.

Practice crimping a couple of terminals without any wire in them. Place the terminal in the proper crimp position for its size. The crimping tab should press on the terminal from the non-seam side of the crimp tube. Figure 7-5 shows how the terminal should be oriented — so the crimp tab is in the middle of the barrel.

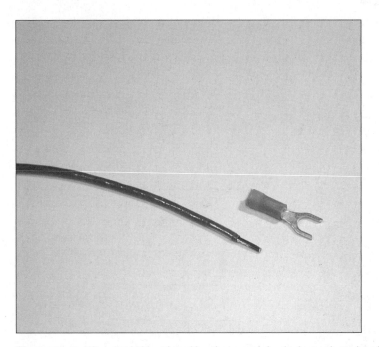

Figure 7-4: Wire should be stripped just far enough for the bare wire to barely protrude from the barrel of the terminal. If the wire and terminal sizes are matched, the wire insulation will not enter the barrel.

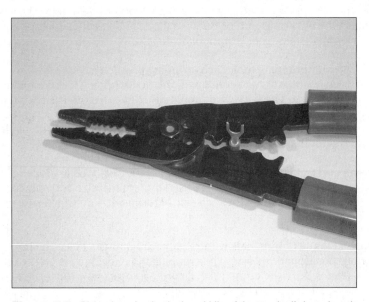

Figure 7-5: Place the crimping in the middle of the terminal's barrel on the opposite side of the barrel's seam.

Now squeeze! As soon as the crimp tab punches into the crimp tube, you can let up. You don't have to strangle the tool! Use a magnifier to look into the crimp tube so you can see the indentation that captures the wire. Look at the back of the terminal to see the clean dimple in the insulation surface (as shown in Figure 7-6). If you didn't make a clean crimp, try again until you feel comfortable with the tool and terminals.

Dimple

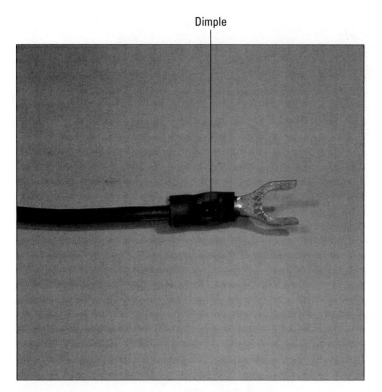

Figure 7-6: The crimped-on terminal should have a nice clean dimple in the non-seam surface of the barrel.

The modular-plug crimping tool

Used to install telephone connectors on what the telephone industry calls "modular cable", the modular-plug crimping tool (shown in Figure 7-7) is a cousin to the tool used for crimp terminals shown in Figure 7-2. It's used in fundamentally the same way: You strip the cable, place wires in the connector, insert the connector into the tool, and squeeze. As the handle is squeezed, a row of tabs press on the modular plug's contacts, forcing their sharp teeth through the insulation and into the wire itself. This connector design is engineered to be simple, inexpensive, and surprisingly strong.

Telephone modular cables and handset cables have two or four wires surrounded by a flat plastic jacket so all the wires are in a row. Connectors that go into a telephone wall jack and the phone itself have the model number RJ-11. They generally have four contacts, although you will occasionally encounter 6-conductor versions that support three separate telephone lines (see the sidebar "Making the Right Call on Telephone Cables"). The cords that connect handsets to phones have a narrower plug called an RJ-10 or RJ-H. A drawing showing the workings of modular connectors is shown in Figure 7-8.

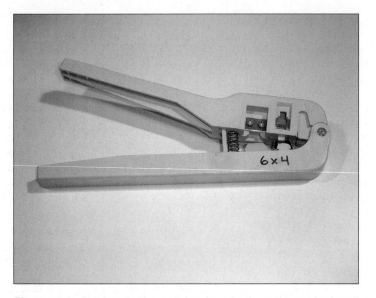

Figure 7-7: This inexpensive modular-plug crimping tool is used to install modular plugs on telephone modular cable and handset cords.

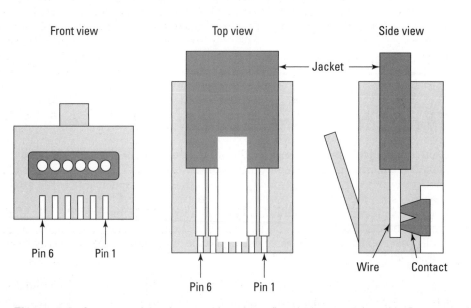

Front view   Top view   Side view

Jacket

Pin 6   Pin 1

Pin 6   Pin 1

Wire   Contact

Figure 7-8: Common modular plugs are crimped onto flat telephone modular cable. Be sure to orient the cable so the right color wire is connected to the right plug contact.

Making the right call on telephone cables

Most telephone-to-wall jack modular cable (called *station cable*) has four wires — but there are common variations that you need to look for when making new cables or replacing old ones. Each telephone line requires only two wires to work. If your home has single-line service, the center two contacts of the modular plug carry your conversations. A second line uses the outer two contacts of the modular plug. Two-line installations are common enough that nearly all phone-to-wall cables are made with four-wire cable and four-contact RJ-11 plugs.

If you look closely at an RJ-11, you will see that it has positions for two more contacts, one on each side of the central four. If two more contacts are added to these positions, the plug becomes an RJ-12, used for three-line installations. That's why an RJ-11 is often referred to as a 6-4 plug; it has 6 positions, 4 contacts An RJ-12 is a 6-6 plug: 6 positions and 6 contacts. RJ-10 handset plugs have only 4 contact positions, making this type a 4-4 plug.

When replacing a cable, look carefully at how many wires it contains — and replace it exactly or the phone may not work properly. If you're making a new cable and aren't sure

how many lines the phone requires, look into the wall jack and count the number of contacts. Use plugs and cable with the same number of wires. (A 4-wire cable plugged into a 6-wire plug will work for lines 1 and 2, but line 3 won't work.) Table 7-2 shows the old and new color codes for telephone cable. UTP stands for Unshielded Twisted-Pair and is the new standard for both telephone and data cables.

Time for a short history lesson: Tip and ring are old words that go back to the early days of the telephone system. You may have seen photographs of operators sitting in front of large bays of cables and sockets. The cables were called *patch cords* and each was terminated in a plug just like the ¼" stereo phone plugs on headphones today — that's why they're called *phone plugs*. The contact at the very end of the phone plug is called the *tip* and the contact just behind the tip is the *ring*. The larger contact between the ring and plug body is the *sleeve*. The Web site `inventors.about.com/library/inventors/bltelephone7.htm` has a lot more information about the telephone system's history and technology.

Table 7-2:	Modular Cable Color Codes			
Connector pin number	RJ-11 function	RJ-12 function	Old color code	**New (UTP) color code
1	None	Line 3, tip	*White	*White/green
2	Line 2, tip	Line 2, tip	Black	White/orange
3	Line 1, ring	Line 1, ring	Red	Blue/white
4	Line 1, tip	Line 1, tip	Green	White/blue
5	Line 2, ring	Line 2, ring	Yellow	Orange/white
6	None	Line 3, ring	*Blue	Green/white

* Not present in 4-wire cables.
** Second color is the stripe color.

The RJ-45 crimping tool

The RJ-45 plugs used for computer network cables are similar to telephone modular plugs and have eight contacts instead of four or six. The crimping tool for RJ-45 plugs is very similar to a telephone modular crimping tool. Figure 7-9 shows RJ-45 plugs and a suitable crimping tool. I prefer the metal crimping tool because more contacts are being pressed into the wire at once. Lightweight, plastic crimping tools will work if you pay a little extra attention to detail.

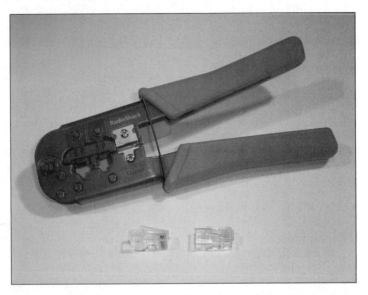

Figure 7-9: Similar to telephone modular connections, the RJ-45 modular plug is used for computer network cable. The crimping tool is similar to that used for the smaller telephone connectors.

Determine whether you need a straight-through or a crossover cable:

- If you're connecting a computer to a network device such as a hub, switch, router, or DSL gateway, you will almost always use a straight-through cable. Check the equipment manual to be sure.

- If you're connecting two computers directly (or a computer directly to some other device that's made to be connected to a hub or switch), then you need a crossover cable.

If you use the wrong type of cable, transmitting circuits will be connected to transmitters and receivers to receivers. (Oops.) This won't cause electrical damage, but no one will be listening or talking correctly and the connection won't work.

Network cable looks like telephone modular cable, but it's not the same. Network cable is a special type called CAT5 UTP (Unshielded Twisted Pair) in which there are four pairs of wires, each pair twisted together separately from the others. Each network signal is carried by one pair of wires. Table 7-3 shows the standard color code for network cables. A discussion of network connector wiring with color pictures is available online at www.incentre.net/incentre/frame/ethernet.html.

There are two ways to make a straight-through cable; use the 568A color code on each end or use the 568B color code on each end. Either will work, but you must use the SAME color code on EACH end.

There is only one way to make a cross-over cable: use the 568A color code on one end and the 568B color code on the other end.

Table 7-3	Network Cable Color Codes	
Pin number	*568A Wire color**	*568B Wire color***
1	White/green	White/orange
2	Solid green	Solid orange
3	White/orange	White/green
4	Solid blue	Solid blue
5	White/blue	White/blue
6	Solid orange	Solid green
7	White/brown	White/brown
8	Solid brown	Solid brown

* - second color is the stripe color

** - for crossover cables, connect the other end using the straight-through color code

Making RS-232 Connectors

A COM port on your PC is an RS-232 serial data interface. Most desktop PCs have one or two, while a laptop PC will have just one at most. The predecessor to USB and Firewire, RS-232 was the standard connection between computers and modems and data terminals for many years. Even though it's being replaced by faster and less expensive technology, RS-232 interfaces are still very common. The task "Making an RS-232 Data Cable" later in this chapter shows you how to connect a 9-pin standard version of the RS-232 connector.

Figure 7-10 shows examples of the two types of DB-style connectors used with RS-232 connections. (There are other DB-style connectors, but they're not often encountered.) The DB-9P has 9 male pins and the DB-25P has 25 male pins. DB-9S and DB-25S connectors have female sockets instead of pins. Connections to the pins can take three forms:

- **Solder cup.** This is the one you're most likely to use in the task later in this chapter.

- **Ribbon-cable crimp.** This is also known as *insulation displacement connection* or IDC.

- **Individual crimp pin/socket.**

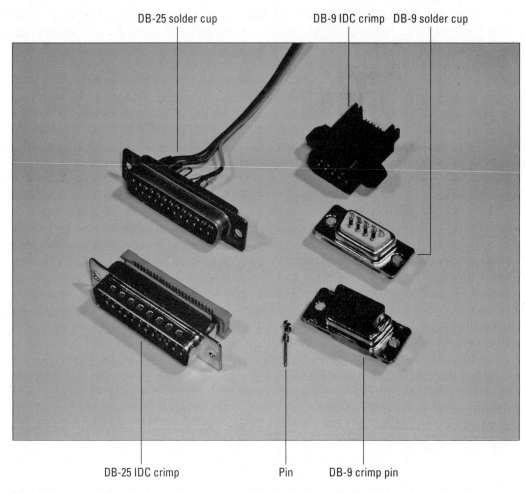

DB-25 solder cup DB-9 IDC crimp DB-9 solder cup

DB-25 IDC crimp Pin DB-9 crimp pin

Figure 7-10: The DB-9 and DB-25 connectors are the standard for RS-232 interfaces.

The RS-232 interface has two data lines, RxD (receive data) and TxD (transmit data), a ground connection, and a set of control signals to manage the flow of data between the devices. Table 7-4 shows the common control signals and the pins to which they are assigned on DB-9P and DB-25P connectors. (DB-9S and DB-25S have slightly different pin assignments.) You'll need a magnifier and good lighting to locate the pin numbers on the back of the connectors. Most equipment doesn't use many of the control signals — and may not use any — even though the RS-232 cables you buy in the store may have every pin connected.

Table 7-4	RS-232 Signal and Pin Assignments	
Signal	*DB-9P Pin*	*DB-25P Pin*
Signal Ground (GND)	5	7
Transmitted Data (TxD)	3	2
Received Data (RxD)	2	3
Data Terminal Ready (DTR)	4	20
Data Set Ready (DSR)	6	6
Request To Send (RTS)	7	4
Clear To Send (CTS)	8	5
Carrier Detect (DCD)	1	8
Ring Indicator (RI)	9	22

There are two common configurations of data and control signals:

- ✔ **3-wire RS-232:** This is the simplest, consisting of GND, RxD, and TxD. No control signals are used and the data flow is managed by software in each device.

- ✔ **5-wire RS-232:** This configuration uses the 3-wire signals plus Request To Send (RTS) and Clear To Send (CTS). No, there won't be a quiz on this! But you do need to know what to look for in the equipment manuals to tell you how many wires to use.

In computer networking, a cable that allows two computers to talk directly without any intervening network equipment is called a *crossover* cable. (See the task, "Making a Computer Network Cable.") A cable that provides the same function between RS-232 interfaces is called a *null modem* cable because it replaces the pair of modems that would normally be used for computers to talk to each other using RS-232.

Installing a Crimp Terminal

Buy a bag of fork, ring, or spade terminals from an automotive parts or electronics store. These stores will also have combination stripper-crimpers for a few dollars. Make sure the terminals are the right size for the wire you have at home or buy some wire to go with the terminals.

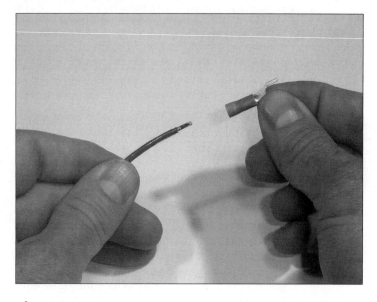

1. Strip a piece of wire about $\frac{5}{16}$. Insert the stripped wire into the crimp tube until the insulation hits the end of the crimp tube. The stripped end of the wire should just be visible at the other end of the crimp tube, extending no more than $\frac{1}{16}$.

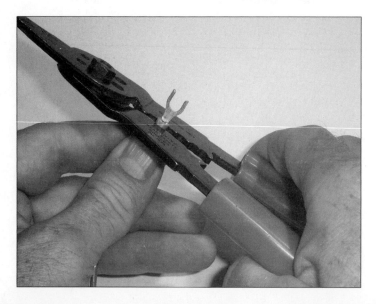

2. Holding the wire in one hand, open the crimping tool with your other hand, and guide the terminal into the proper crimp position. Remember to keep the terminal oriented properly for the crimp tab.

3. With the wire in the terminal and the terminal in the jaws of the crimper, do one last inspection and when you're ready, give the crimping tool a firm squeeze until you feel the crimp tab "hit bottom" and stop moving. Release the handles of the crimper and remove the terminal.

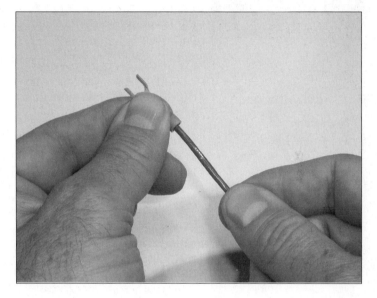

4. The terminal should be firmly attached to the wire. Give the wire a good yank to confirm that it won't pull out. If it does pull out, you need to make a better crimp. Check the placement of the crimping tab and whether you had enough wire stripped and in the crimp tube.

TIP

Practice using different sizes of wires and terminals until you can install a terminal quickly and properly.

TROUBLESHOOTING

If the wire pulls out of the terminal — or seems to move around in the terminal when you pull or twist it — then the terminal wasn't positioned correctly in the crimping tool's jaws and needs to be replaced. Be sure that the crimp tab is positioned over the barrel and not just over insulation. Also be sure that you're not crimping on or near the seam of the barrel (see Figure 7-3). If you find that the electrical connection is intermittent or non-existent, you may be pushing some of the wire insulation into the barrel, preventing the crimp from contacting the wire. Crimp terminals can't be re-crimped, so cut off the bad one and put on a new terminal.

Adding a DC Power Fuse

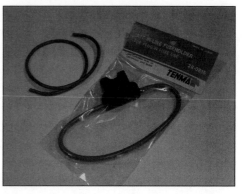

Whenever you connect electronic equipment to a source of DC power, you must consider protecting the power source and the equipment from fire hazards caused by excessive current. If the power source is capable of supplying more than a few watts of power, add a fuse in the equipment power leads. Batteries, particularly vehicle batteries, and large DC power supplies, can cause significant damage in the event of a short-circuit or circuit failure! (The author knows this from personal experience . . .)

If your equipment doesn't already have an internal fuse, it's easy to add one externally by using an in-line DC power fuse kit. In-line fuse kits are used to add a fuse to power leads for equipment that doesn't have an internal fuse (like the one shown in the following figure). Kits are available from electronic parts suppliers and auto parts stores. Fuse holders are available for either glass-cartridge fuses or automotive blade-style fuses. Be sure that the kit is rated to handle the amount of current you need.

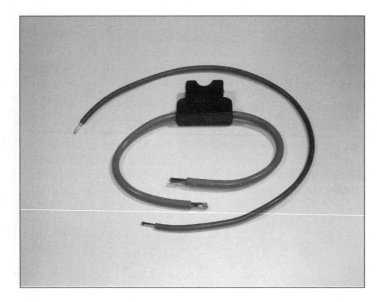

1. The fuse kit is manufactured with a loop of wire connecting the two terminals of the fuse holder. Cut the fuse kit wire in the middle of the loop (or wherever it's most convenient for your installation). Strip the ends of the fuse kit wires and the power lead wires. (The coil of wire in the figure represents the equipment power leads in this task and is not part of the kit.)

2. Install a butt-splice terminal, joining each stripped end of the power leads and the fuse kit's leads. Make sure that the terminal is solidly installed, as described in the previous task.

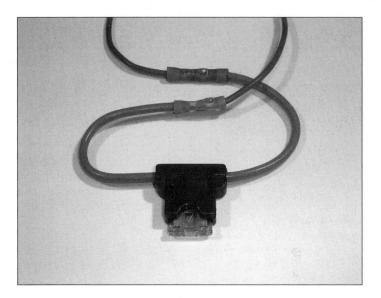

3. Install a fuse in the fuse holder. Done!

If you get the fuse holder installed, but power isn't getting to your equipment, double-check the fuse itself. Use a magnifying glass to inspect the fuse or measure its resistance from end-to-end with a meter (see Part IV for help). If the fuse is good, check the butt-splice crimps. Finally, make sure the fuse holder is assembled properly.

Choosing a fuse

There are many sizes, types, and styles of fuses — how do you choose one? For electronics powered from DC voltages of 24V or less, blade-style automotive fuses are a good choice. They are widely available from both electronics suppliers and at auto parts stores, as are the inline fuse kits. The fuses are available in three sizes; mini, ATO (the most common), and maxi, with ratings of 1 to 80 amps. Glass-cartridge fuses (see Chapter 9) are best for AC power circuits, but are found in older vehicles as well.

The current rating of a fuse is the maximum current the fuse will conduct without opening or "blowing." The greater the overload, the faster the fuse will blow. *Fast-blow* fuses open quickly when the rating is exceeded.

Slow-blow fuses carry current at or near their rated current for some time before opening. To prevent fire hazards from short circuits, pick a fast-blow fuse with a current rating of two to five times the highest current the equipment will draw under normal operating conditions. If the goal of using the fuse is to protect the equipment from drawing too much current when it's overloaded for a period of time, then pick a slow-blow fuse with a current rating closer to the maximum normal current draw.

You can find detailed information on fuses, called *fuseology*, at this Web site:

www.circuitprotection.ca/
fuse-ology.html

Installing a Telephone Plug

Isn't it amazing how often telephone cords are so often just inches too short? In this task, you get a crack at repairing or making your own cords — as long as you want! The same procedure can be used for telephone handset cords.

1. If you're repairing a bad cable, cut off the plug at the end, leaving a clean end with the cut directly across the cable. For new cables, just be sure that the end of the cable is a clean, square cut.

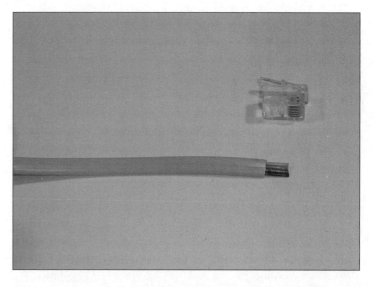

2. Use the strippers built into the modular crimping tool to strip the cable as follows: Insert the cable into the strippers until it contacts the built-in depth gauge stop. Squeeze the handles — and while you're holding firm, pull the wire out of the strippers. The jacket will come off and you'll be left with four (or six) wires in a row.

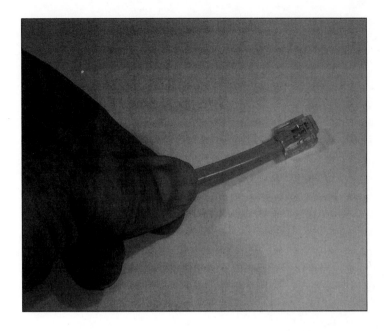

3. Insert the stripped end of the cable into the plug, taking care that the wires go in the positions with contacts. Double-check to make sure you have the right colored wire in the right position. Looking down on the side of the connector with the locking tab as show, the colors should read (left to right) black-red-green-yellow. The jacket of the cable should extend into the back of the plug.

4. Insert the plug (with cable inserted) into the modular crimping tool until the locking tab on the plug snaps out, just as it does when you're plugging the cable into a wall jack.

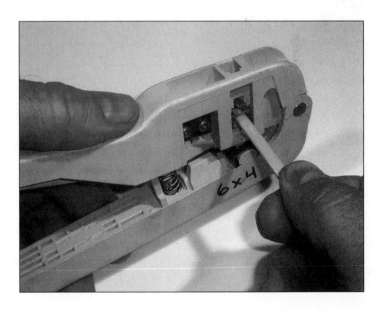

5. Squeeze the handles on the tool until they stop moving together. This doesn't require much force; a firm squeeze will do the job.

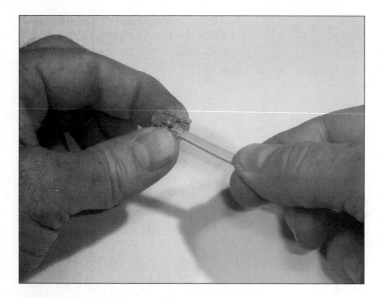

6. Remove the plug from the crimping tool by pushing the locking tab toward the cable and then pulling the plug out of the tool. Give the plug a close visual inspection. You should see the teeth of the end contacts disappearing into the wire insulation. Give the cable a light pull to be sure it won't come out of the plug. If you're making a complete cable, repeat these steps for the other end of the cable. Otherwise try the cable with the phone!

If the phone doesn't work with the new cable, double-check your color code to be sure the right wire is at the right contact position. Do a sanity check by replacing the new cable with a *known* good cable, if possible. If the phone doesn't work with that cable either, the problem may be in the wall jack, which is described in the next task.

Replacing a Telephone Jack

Stuff You Need to Know

Toolbox:

✔ Small, flat-bladed screwdriver
✔ Wire cutters

Materials:

✔ RJ-11 modular wall jack

Time Needed:

Less than an hour

Sometimes telephone wall jacks are damaged and need to be replaced. If you're remodeling or updating phone wiring, you will eventually have to replace the old-style telephone's wall-mounted screw-terminal blocks with modular jacks. This task shows you how to install the new jack — whether a replacement or a completely new jack.)

The following figure shows both the old and new type of telephone wall jack. An old-style telephone wall terminal requires the phone cable to be permanently attached to screw terminals. These should be replaced with new modular jacks. New jacks consist of a connector in the cover that snaps or screws on to a terminal block mounted on the wall. For more information about telephone wiring, take a look at the previous task. The Web site www.homephonewiring.com also has a lot of good information about your home's phone-wiring system.

Old style   New style

1. Start by finding the telephone service box for your house, opening its cover, and removing the modular plug as shown, disconnecting the line. Disconnect the line first will let you work on the wiring without any unusual electrical activity on your line or lines. If you're uncomfortable accessing your service box, the phone company will disconnect and reconnect for you, for a fee.

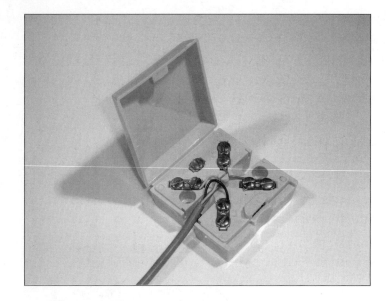

2. Unscrew the cover from the old block and remove the phone line cable (called *premises wiring* by the phone company) from all screws. Keep the wires apart. There are no dangerous voltages present, but shorting the wires together may cause the phone company to think you're dialing if you're working on a live line. If the old block has wires held between metal fingers, that is a *punch-down* block — just cut the wires at the fingers and re-strip them.

3. Remove the terminal block from the wall and replace it with the new jack's terminal block. If the screw hole is loose or has been damaged, use a plastic wall anchor to hold the new block firmly.

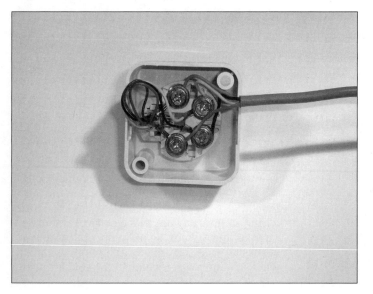

4. Reattach the premises wiring to the new jack's terminals, being careful to match the wire colors and the terminal colors exactly.

5. Reattach the cover of the new jack to the terminal block.

6. Return to the telephone service box, reconnect your phone line, and secure the cover. Now plug a phone into the new jack and make sure it works!

If you're a renter, the rental agreement may restrict what you can do to your home's phone wiring. Check first!

If a phone doesn't work with the new jack, did you remember to reconnect the line at the service box? If so, double-check the terminal wiring to be sure the premises wire colors match those from the jack connector. If you were working on a live line, you may have confused the phone company's equipment as you disconnected and reconnected the wires. Wait for a few minutes to reset the line and try again. Try a different phone or change the phone-to-wall cable. If the phone receives a dial tone, but won't dial, the tip and ring connections (see the previous task) are probably reversed. If the phone is still completely dead, telephone-line testers are available from home-improvement stores and many electronics vendors. If all else fails, the phone company will come to your home and troubleshoot the problem (for a fee).

Making a Computer Network Cable

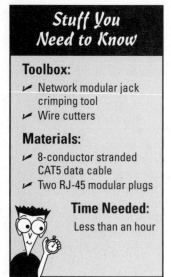

Stuff You Need to Know

Toolbox:

✔ Network modular jack crimping tool
✔ Wire cutters

Materials:

✔ 8-conductor stranded CAT5 data cable
✔ Two RJ-45 modular plugs

Time Needed:

Less than an hour

Start by reading through the sections earlier in this chapter on the modular crimping tool as well as the previous task, "Installing a Telephone Plug" to familiarize yourself with modular cable, connectors, and crimping tools. A computer network cable is a kind of big brother to telephone modular cables. In this task you'll learn how to make both kinds of computer network cables; *straight-through* and *crossover*.

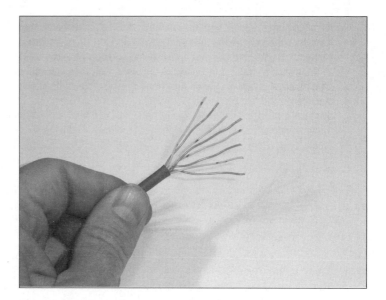

1. Strip about 1" of the modular cable jacket using the crimping tool (if yours has a cable stripper) or by scoring the jacket with a sharp knife (don't nick the wire insulation). Untwist and straighten all the wires and arrange them in order of the straight-through column in Table 7-3.

TROUBLESHOOTING

If the cable doesn't seem to work, give each plug a very close inspection, looking for both color code and complete crimping. Make sure that the pin-by-pin color code is correct. Each wire should be completely inserted into the plug. If there is any question, cut the connector off and start over. Cable checkers are available at computer-supply stores and electronic vendors, but such a tool may not be a good investment unless you're planning on making a lot of cables. Computer shops can usually check cables, too.

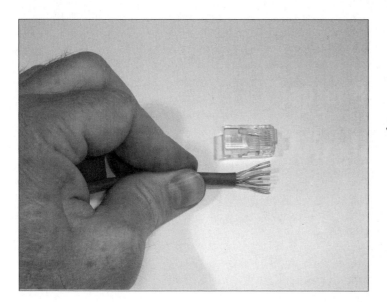

2. Trim the wires square with the end of the jacket and long enough (about ½″) so when the cable is inserted into the RJ-45 body, the cable jacket is also inside the plug's body.

3. Insert the wires into the RJ-45, making sure the wire order is correct. Seat the cable firmly to be sure all wires are all completely inserted. It may help to look at another network cable to confirm that you're doing it right.

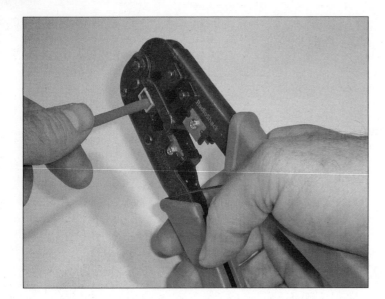

4. Insert the plug into the crimping tool and squeeze the handles until they meet or hit a stop.

5. Remove the plug from the crimping tool by pushing the locking tab towards the cable and pulling the plug out of the tool. Give the plug a close visual inspection. You should see the teeth of the end contacts disappearing into the wire insulation. Give the cable a light pull to be sure it won't come out of the plug.

6. Make the other end of the cable in the same way. Double-check to be sure that you are looking at the connector in the same orientation on both ends of the cable and that you've positively identified pin 1 on both according to the drawings. If you're making a straight-through cable, arrange the wires in the plug using the straight-through column of Table 7-3. If you're making a crossover cable, use the crossover column of Table 7-3.

Making an RS-232 Data Cable

Stuff You Need to Know

Toolbox:

- ✔ Soldering iron and solder
- ✔ Wire cutters
- ✔ Wire strippers or sharp knife
- ✔ Small needle-nose pliers
- ✔ Small vise

Materials:

- ✔ Multiconductor data cable (at least three wires)
- ✔ Short length of hookup or bare wire
- ✔ Heat-shrink tubing to fit cable
- ✔ DB-9 connector

 Time Needed:
Less than an hour

In this task, you're going to make a 3-wire DB-9P RS-232 cable and add jumpers between two sets of the four control signals; RTS to CTS (pin 7 to 8) and DSR to DTR (pin 4 to 6). This is a common trick used by manufacturers to let software do the heavy lifting of data-flow control while at the same time keeping the electronic circuitry happy. For a LOT more information on RS-232, read the Wikipedia entry at `http://en.wikipedia.org/wiki/RS-232`.

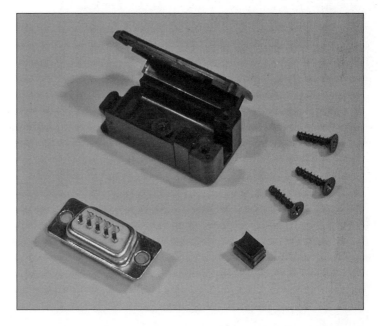

1. Disassemble the connector or unpack the connector kit.

2. Use a magnifying glass to identify which pin is which; put a pencil mark by the solder cup for Pin 1 to remind yourself of the correct orientation.

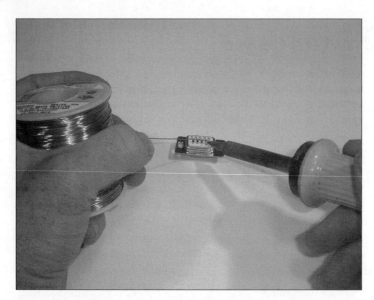

3. Tin each solder cup that you'll be using; Pins 2 through 8.

4. Strip the jacket of the data cable back about 1". When the wires are soldered to the connector terminals, the jacket should be inside the shell of the connector.

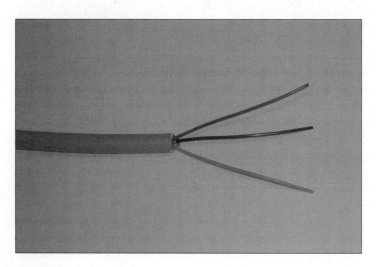

5. If the cable has more than three wires, snip off the extras at the jacket, including any shield wires.

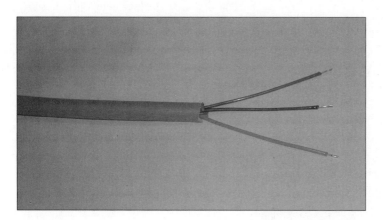

6. Strip each remaining wire about ⅛"
and tin the end. Strip and tin two 1"
pieces of hookup or bare wire, too.

7. Place the connector in your vise with the solder cups facing up.

8. Take one of the 1" pieces of wire
and heat the solder cup for Pin 4
with the soldering iron until the
solder melts, then use the needle-
nose pliers insert the tinned end of
the wire into the cup and melt that
solder, too. Remove the iron while
holding the wire steady until the
solder solidifies. Repeat for the
other end of the wire and Pin 6.

9. Repeat Step 8 for the remaining 1" piece of wire and Pins 7 and 8.

10. Since this is just a practice connector, it won't matter which of the cable wires is connected to which pin. (When you make a real connector, you'll have to know exactly which pin carries which signal — information provided by the equipment manufacturer.) Repeat Step 8 for each of the three cable wires and Pins 2, 3, and 5.

11. Install the shell of the connector, being sure to capture the jacket in the cable clamping area at the back of the shell.

Sometimes you need a cable with a DB-9 or DB-25 on one end and free wire ends on the other for attachment to a terminal strip or soldering to a printed-circuit board. Save yourself a lot of work by purchasing a commercially-made cable with DB connectors attached — get a length that's twice as long as you need. Then cut it in half and use a continuity tester or multimeter to figure out which wire is connected to which pin. You're done — and you've got a spare cable to boot!

Chapter 8

Wiring for Wireless Radio

Tasks Performed in This Chapter

- ✔ Learning how to use a coaxial connector crimping tool
- ✔ Installing a TV-style connector
- ✔ Installing a CB-style connector
- ✔ Installing a scanner-style connector
- ✔ Waterproofing an exterior connector

All the tasks in this chapter involve putting connectors on a piece of coaxial cable to be used with some kind of *radio-frequency* (RF) signal — TV, citizens band (CB), or amateur (ham) radio, or a scanning receiver. These connectors and the cable are a bit special because they have to carry signals that are far higher in frequency than audio and power. Radio receivers have to work with signals that are incredibly weak — a billionth-of-a-watt signal is considered very strong! CB and amateur transmitters depend on the connector being just right so they can deliver all that precious radio power to the antenna. A short bit of background on coaxial cable will help you understand why special care is needed for radio frequency connectors.

The Case for Coaxial Cable

Coaxial cable (or just "coax" for short) has two *conductors* that share a common central axis; they are *co-axial*. You can see this for yourself by looking at the end of a piece of coaxial cable. (Go ahead, I'll wait!) See? Figure 8-1 shows the parts of the cable a little more clearly. The signal carried by coaxial cable flows on the center conductor and on the *inside* of the outer shield. None of the internal signal is intended to flow on the outside of the shield; here the flow is similar to water in a hose. Coax is used for routing RF signals around your house or in the car because it protects the RF signals from interference and noise. No matter what contacts the outside of the shield, the signals inside are unperturbed.

Conductor: In a cable, a conductor is any metal wire, foil, or woven braid that carries a signal. In a cable with multiple conductors (known as *multi-conductor*), each conductor is electrically insulated from all the other conductors. Coaxial cable has two conductors (the center conductor and the shield) that share a common central axis.

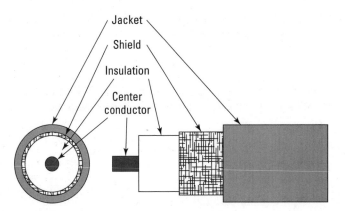

Figure 8-1: In coaxial cable, two conductors — the center conductor and the shield — share a common central axis.

Frequency: In alternating current, electrons move backward and forward in a regular pattern known as a *cycle*. Each cycle comprises one back-and-forth movement. Frequency is a measure of how cycles occur every second. Frequency is measured in Hertz (Hz), which is the same as a cycle per second. Different units of frequency are abbreviated kHz (kilohertz), MHz (megahertz), and GHz (gigahertz).

Radio frequency (RF): RF signals have frequencies higher than 30,000 Hz. AM radio signals have frequencies of 550 to 1700 kHz (kilohertz). FM signals range from 88 to 108 MHz (megahertz).

All coaxial cable are not the same, even though they all have the same arrangement of center conductor, insulation, and shield. Each type of coax has a *characteristic impedance,* usually 50 or 75 ohms. This value relates to how the RF signals flow in the cable; it's *not* the resistance you might measure with your multimeter from end to end. TV and video systems have standardized 75-ohm cable such as RG-6, RG-11, and RG-59. Radio systems usually use 50-ohm cables such as RG-8, RG-58, and RG-213. Using the right cable is important in getting the best system performance.

As you might have guessed, protecting the integrity of that outer shield is really, really important. Any loss of integrity in a connector for RF signals generally means poor or no reception, lots of noise and interference, and unhappy transmitters. That's why it's also really, really important to install *coax connectors* properly. The crimp-on connectors and crimping tool make that process fairly easy, as long as you follow instructions and don't cut corners.

Using a Coaxial Connector Crimping Tool

The coaxial connector crimping tool (which we'll call a coax crimper to save our breath) in Figure 8-2 is heavier and stronger than the plastic crimpers used for telephone cable. It has to be to crimp metal rings and tubes!

This type of coax crimping tool uses a *compound ratchet* action to multiply your hand's squeezing force. Between the blue handles you can see the teeth of the ratchet. Using a ratcheting action ensures a consistent crimp, even if it takes a couple of squeezes to get the job done. If you get part of the way through a crimp and decide that you need to stop crimping and release the connector, look for the ratchet-release lever just behind the ratchet cam. Push this lever forward to release the ratchet and open the crimper's jaws. When you have squeezed the crimper completely shut, the tool automatically releases the ratchet and opens its jaws. Give this a try with your own coax crimper. You'll hear the clickety-click as the jaws come together, then a pause, followed by one or two more clicks as the automatic release occurs.

The business end of the crimper is its jaws. The jaws hold a pair of complementary *dies* that have hexagonal holes for the connector pieces. The crimper's instructions will show you which hole to use for which connector and cable type. The large holes are used to crimp a metal crimping ring over the cable's shield. The small holes crimp the connector's center pin (if there is one). When you purchase a coax crimper, be sure to get a model that comes with the proper dies for connectors and the type of cable you'll be working with.

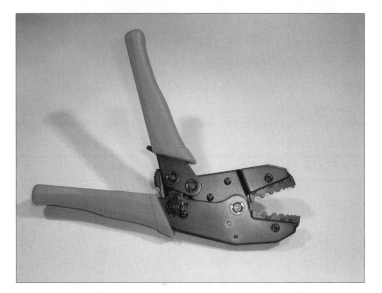

Figure 8-2: The coaxial connector crimping tool compresses the pieces of the coax connector to make a consistent that maintains its integrity at radio frequencies.

Coax Connectors — All in the Family

Because radio signals come in so many forms, frequencies, and strengths, one or two types of connector just can't do all the jobs properly. The different types of connectors that have been created to do those jobs are grouped into *families*. Within each family, there are plugs, receptacles (also called *jacks* or *sockets*), and *adapters* — connectors that mate plug to plug, receptacle to receptacle, and even allow connections between connector families. In most electronic work, you only encounter the three connector families described in this chapter: Type F, UHF, and BNC. Examples of connectors and adapters from these families are shown in Figure 8-3. Table 8-1 lists the common names or designators for the most common connectors in each family.

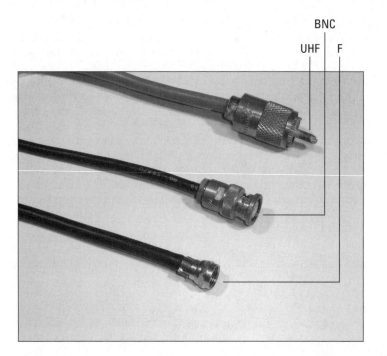

Figure 8-3: Coax connectors of the three most common connector families: Type F, UHF, and BNC.

Table 8-1	Coax Connector and Adapter Designators		
Connector Type	*Type F*	*UHF*	*BNC*
Plug (Male, RG-6/RG-8X/ RG-59 cable)	Plug (listed by cable type)	PL-259 plus UG-176 reducing adaptor	UG-260B/U
Plug (Male, RG-58 cable)	Plug (listed by cable type)	PL-259 plus UG-175 reducing adaptor	UG-88B/U
Plug (Male, RG-8/ RG-213 cable)	Plug (listed by cable type)	PL-259, no reducing adaptor	UG-959/U
Receptaple (Female)	Panel	SO-239	UG-1094/A
Double female or barrel	Barrel or double female	PL-258	UG-914S
Double male	Male-to-male	PL-269	UG-491A
Bulkhead (routes connections through panels or walls)	Bulkhead or double female	PL-224, PL-363	UG-492A

Type F connectors

Type F connectors are very inexpensive and easy to install. You have seen them on the ends of the cable that connects your TV to a cable TV wall outlet. They are designed for coaxial cables that have a solid center conductor. To make installation really easy, these connectors don't even have a center pin — they use the cable's center conductor instead! The body of the plug has a threaded ring called a *shell* that screws on to receptacles and female adapters.

UHF connectors

Like Type F connectors, the UHF connector family is threaded, meaning they attach to each other using threaded shells. (In this case, the name *UHF* is not related to the Ultra High-Frequency band of TV signals.) The shells of UHF connectors are separate from the connector body. UHF connectors are widely used for CB and amateur radio, shortwave broadcast receivers, marine radios used on boats, and commercial radios.

The stock version of the PL-259 plug is made to be fitted to RG-8 coaxial cable or one of the same diameter, such as RG-213. To be used with the smaller coax that's more common with CBs and mobile radios (RG-8X or RG-58) a reducing adapter must be screwed into the back of the plug.

BNC connectors

The BNC connector family of connectors is found on scanners and other receivers and is also common on electronic test equipment. It is a *bayonet*-style locking connector. (The UHF and Type F connectors are threaded.) To connect a BNC plug to a BNC receptacle, align the bayonet pins of the receptacle with the slots of the plug, slide the connectors together and twist clockwise until the pins slide into the locking position, called a *detent*. To unplug, press the connectors together, freeing the pins, and twist counterclockwise.

BNC connectors are used primarily with smaller coaxial cables; RG-6, RG-58, and RG-59. Remember to purchase the BNC connector designed to fit your cable. Your coax crimper must also fit the cable. As with UHF connectors, a traditional style of connector is available that uses solder to attach the center pin and a screw-in compression clamp to connect the braid. These work well, but they take a lot of practice to install correctly. Compression-style connectors similar to Type F compression connectors are available, as well. As a beginner, stay with the crimp or compression type connector.

Adapters are available for almost any combination of plugs and receptacles imaginable. In addition, you can combine or reroute cables with *tee* adapters that join three cables (all-male, all-female, 2-male, 2-female), or *right-angle* adapters that make connections in tight spaces.

Not THAT kind of stripper!

If you expect to be installing a lot of RF connectors, you should spend a few bucks and get a coax stripping tool or coax stripper of the type shown in Figure 8-4. This convenient little wonder makes short work of trimming the cable jacket and making a nice clean cut into the center insulation without nicking the center conductor. What takes a minute or more using a knife or razor blade takes less than ten seconds with this tool. You can buy a coax stripper wherever you buy your coaxial cable and coax crimper.

Figure 8-4: A coax stripper "automagically" removes jacket, shield, and center insulation with a few twists — a big time-saver!

Installing a TV-Style Connector

Coaxial cables carrying TV signals, whether from an antenna or a cable TV system, connect to a TV set or DVD player using a Type F connector. There are two types of crimp-on Type F connectors; the original crimp connectors with a separate crimp ring and the direct crimp connectors with the inner and outer crimp surfaces an integral part of the connector. This task shows the direct crimp style, but the same instructions can be used for both types as long as you remember to slide the first type of connector's crimp ring onto the cable first!

It is important to get the type of connector that matches your coaxial cable. When in doubt, ask a salesperson. The variations between connector types are small and hard to judge by eye. The packaging or description of the connector will state what cable types the connector will fit — RG-59, RG-6, RG-6A, RG-6 tri-shield, RG-6 quad-shield, or RG-11. Don't try to use the wrong type of connector! Even if you can get the connector on the cable, the connection will soon degrade, work loose, and fail.

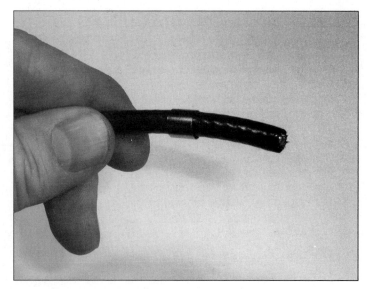

1. Start by cutting the end of the cable square and clean. If your connector has a separate metal ring, slide it on to the cable now.

2. Read the assembly instructions on the connector packaging (or the manufacturer's Web site) to determine how much of the jacket, shield, and center insulation to remove. Different manufacturers and styles of connectors require different amounts of shield (just braid or braid-and-foil) to be exposed. All Type F connectors require a quarter inch of the cable's center conductor to be exposed.

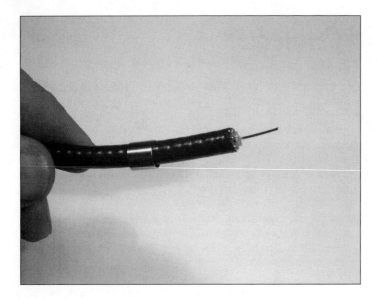

3. Strip the cable according to the instructions, using a coax stripper or very sharp knife. Be sure you don't nick the cable's center conductor. If nicked, the wire will break after flexing once or twice; if you nick the wire, it's best to start over. The end of the cable should look like the figure. (Note: if you are installing connectors on quad-shield cable that has double layers of braid and foil, the stripping process should only remove the outer set of braid and foil.)

4. If required by the instructions, fold the braid back and over the cable jacket. Make sure none of the tiny strands of braid cable are making a short circuit to the center conductor wire. Otherwise, trim the braid and foil flush with the end of the jacket as shown.

5. Push the body of the connector onto the cable without turning it from side to side. The inside sleeve of the connector should go under the braid and jacket. Press against a vise as shown in the figure. Keep pushing until the end of the center insulation is at the base of the connector's front section and the center conductor sticks out through the center hole, not touching it at all and slightly beyond the end of the connector. If your connector has a separate metal ring, slide it back over the braid or jacket, keeping the braid folded back over the cable jacket. The ring should be as close to the connector body as possible.

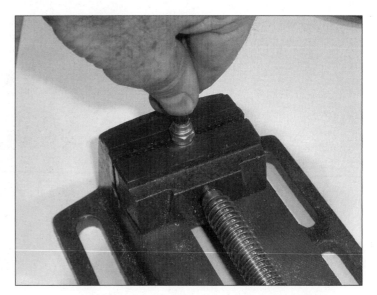

If the connector can be easily pulled off with your fingers, it wasn't installed properly. The most likely reasons are not stripping the cable to the right dimensions or letting the connector slip off the end of the cable as you inserted it into the coax crimper. You can't reuse or recrimp the connectors, so cut it off and start again.

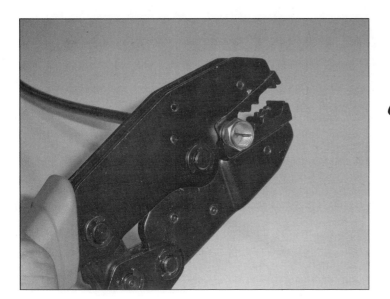

6. Place the entire connector assembly in the appropriate hole of the coax crimper's jaws so the crimp ring is centered within the hole. Don't let the cable slide back out of the connector — keep it seated firmly in the connector!

7. Squeeze the handles of the coax crimper, evenly and firmly, until the ratchet is released. Give the connector a close visual inspection to be sure the center conductor isn't contacting any part of the connector body and is approximately flush with the front surface of the connector. Repeat Steps 1 through 6 if you're making a jumper with connectors on each end.

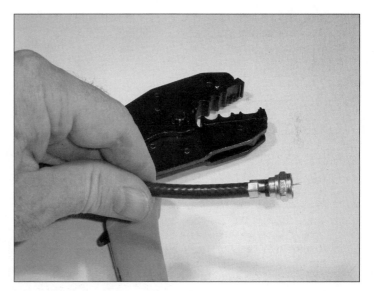

Twist-on and compression connectors

The steps to install a crimp-on connector are mostly repeated for two new styles of Type F connectors: compression and twist-on. Instead of squeezing the braid between inner and outer metal rings, the braid is pushed against the forward surface of the metal connector body. A plastic ring or metal housing is pushed into the back of the connector by another type of crimping tool or by screwing on a threaded shell. These connectors are a bit more expensive, but they have better water resistance for installation outside.

Installing a CB-Style Connector

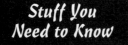

Stuff You Need to Know

Toolbox:
- Coax crimping tool
- Coax stripper or sharp knife
- Wire cutters

Materials:
- Coaxial cable
- PL-259 crimp-on connector for the size of coax you are using (two if you're making a jumper) to fit your coax

Time Needed:
Less than an hour

Just like TV signals, citizens band (CB) radios depend on high-quality connector installation for maximum range on both transmit and receive. CB radios use frequencies near 27 MHz — definitely radio frequencies! This task demonstrates how to install the most common type of connector used on CB radios — the PL-259. The connector attaches to the cable in much the same way as the Type F connectors in the previous task. Make sure you have the right connector for the size of cable being used; RG-8 (or RG-213), RG-8X, RG-59, or RG-58.

The coax crimper used for PL-259 connectors operates exactly the same as the Type F coax crimper used previously. In fact, many coax crimpers can do both Type F and PL-259 crimps — or can be made to by changing the dies in the crimper jaws.

There's a lot more information about how to install and operate amateur and CB radios and antennas in the author's books, *Ham Radio For Dummies* and *Two-Way Radios and Scanners For Dummies* (both published by Wiley Publishing, Inc.).

1. Start by cutting the end of the cable square and clean. Then take a minute to read the manufacturer's instructions for assembling the connector. You can find the instructions on the connector package or on the manufacturer's Web site.

2. If you are using a coax stripping tool, use it to remove the jacket, braid, and center insulator in a single operation; then proceed to Step 5. Otherwise, as seen in the figure, use a sharp knife or razor blade to score and remove the cable jacket so the shield braid is exposed according to the assembly instructions. If you don't have assembly instructions, remove enough jacket so that the braid extends about ⅛" past the end of the connector's hollow tip; make sure the jacket is even with the base of the connector. Be careful to avoid nicking the braid where the jacket is removed.

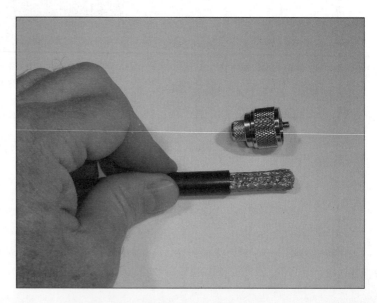

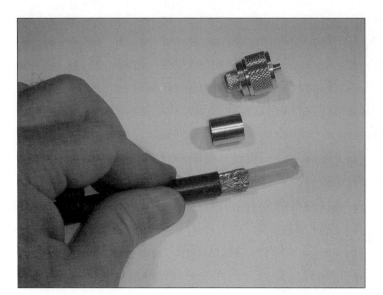

3. Trim the braid according to the instructions or so the remaining length is equal to the length of the crimping ring.

4. Score and remove the center insulator from the end of the braid, exposing the center conductor according to the instructions. If the center conductor is stranded wire, be sure all the strands stay together and do not fray or bend. Be careful to avoid nicking the center conductor.

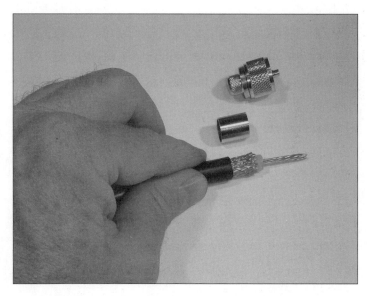

5. If required by the instructions, fold the braid back over the jacket and slide the separate metal crimping ring onto the cable. Otherwise, slide the metal ring over the jacket.

6. Push the body of the connector on the cable so the center conductor goes through the hollow center pin and the inner crimp surface slides under the braid. Keep pushing until the connector body is seated against the braid and jacket. Use a multimeter to check the resistance between the center pin and the shell — it should show an open circuit. If it doesn't, one of the center conductor strands may not have gone into the hollow center pin — pull the cable out, check it, and then reassemble.

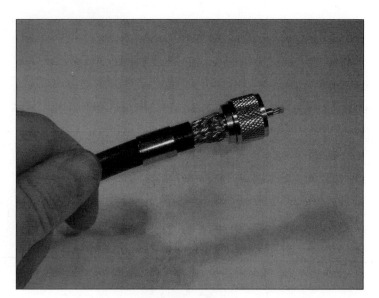

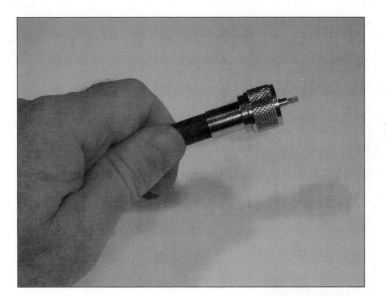

7. Slide the crimping ring over the braid folded back against the jacket until it is touching the connector body.

8. Place the entire connector assembly in the appropriate hole of the coax crimper's jaws so that the crimp ring is centered within the hole. Don't let the cable slide back out of the connector — keep it seated firmly in the connector!

9. Squeeze the handles of the coax crimper evenly and firmly until the ratchet is released.

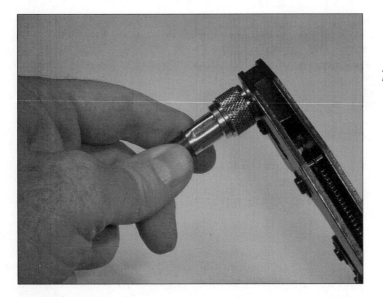

10. Now crimp the center pin of the connector in the same way. (Many people solder the center conductor for good measure after it's been crimped.) Trim any exposed center conductor or braid. Use a multimeter to check the resistance between the center pin and the shell — it should show an open circuit. If it doesn't, you'll have to cut off the connector and try again!

Installing a Scanner-Style Connector

Scanners have become quite popular for everything from keeping up with the local police department to listening in to crews and drivers at the race track. The range of portable and handheld scanners can be greatly improved with an antenna mounted outside and clear of obstructions and metal objects, connected to the scanner with coaxial cable that has a coax connector at each end. This task shows you how to install a BNC plug that can connect to most scanners.

If the length of coax between your scanner and the antenna is significantly more than 50 feet, you may want to consider using larger, lower-loss coax such as RG-8. But the larger cable can be too stiff for convenient attachment to a handheld scanner. The solution is to put UHF plugs on the larger cable and add a UHF plug-to-BNC receptacle adapter at one end. Then make a short (3- to 6-foot) jumper of smaller, more flexible coax, with a BNC plug on each end to connect the scanner to the larger cable.

Some antennas don't have BNC connectors — they may use UHF or some other type of connector. Find out before going to all the trouble of installing a connector on the end of a long run of cable, only to have to remove it!

1. Start by cutting the end of the cable square and clean. Then take a minute to read the manufacturer's instructions for assembling the connector. They will be on the connector package or on the manufacturer's Web site.

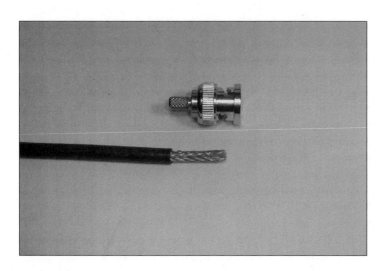

2. If you are using a coax stripping tool, use it to remove the jacket, braid and center insulator in a single operation and proceed to Step 5. Otherwise, use a sharp knife or razor blade to score and remove the cable jacket so that the shield braid is exposed according to the assembly instructions. If you don't have assembly instructions, remove the jacket for a distance equal to the inner base of the connector's insulator to the end of the crimping surface. (¾" for the connector shown in this sequence of figures.) Be careful to avoid nicking the braid where the jacket is removed.

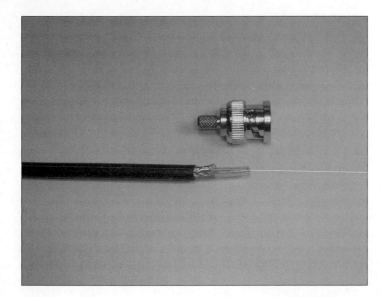

3. Use sharp wire cutters (or a very sharp knife or razor blade) to remove part of the exposed braid according to the assembly instructions, leaving the same length of braid as the crimping surface. On the connector shown, ⁵⁄₁₆″ of braid should be left.

4. Slide the crimping ring (also called a "ferrule") onto the cable. Score the exposed center insulator according to the assembly instructions and remove it. If you don't have assembly instructions, remove the center insulation, leaving ¹⁄₁₆″ extending beyond the end of the braid.

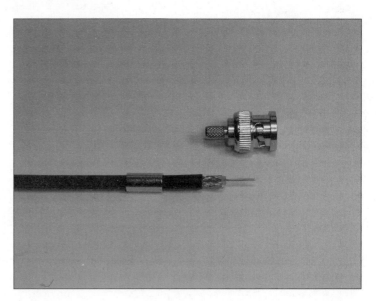

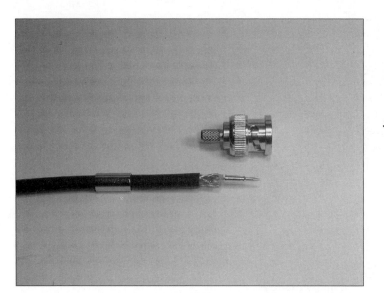

5. Slide the connector's center pin onto the exposed center conductor. If it does not reach the center insulation, trim the center conductor a little and try again.

6. With the center pin resting on the center insulation, insert the pin into the crimper.

7. Crimp the center pin onto the center conductor. Make a close visual inspection to be sure the crimp is solid and that the pin isn't bent.

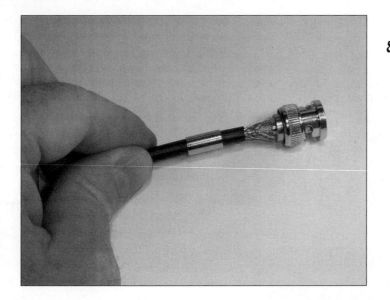

8. Slide the body of the connector over the pin and onto the cable. The inner crimping surface should slide under the exposed braid. Make sure none of the fine braid wires can make a short circuit to the center pin. The center pin should extend just to the forward surface of the connector. There may be a slight snap or click as the cable reaches full insertion. (It is not uncommon to have to remove a small additional amount of jacket to allow the cable to be fully inserted into the body of the connector.)

9. Slide the crimping ring forward until it contacts the rear of the connector body.

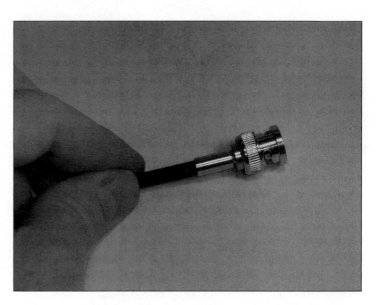

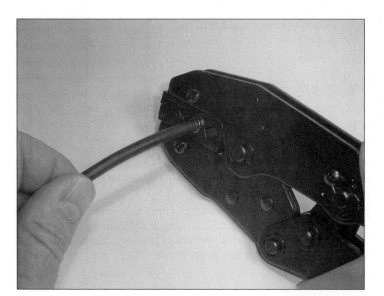

10. Place the entire connector assembly in the appropriate hole of the coax crimper's jaws so the crimp ring is centered within the hole. Don't let the cable slide back out of the connector — keep it seated firmly in the connector!

11. Squeeze the handles of the coax crimper evenly and firmly until the ratchet is released. Remove the connector from the crimper.

12. Use a multimeter to check the resistance between the center pin and the shell — it should show an open circuit. If it doesn't, you'll have to cut off the connector and try again!

13. Repeat Steps 1 through 6 if you're making a jumper with connectors on each end.

If your cable fails to get a signal from one end to the other and the multimeter doesn't show a short circuit, the problem may be with the center pin. If the cable worked a bit loose from the connector before crimping, the center pin may be too deep in the connector to make contact with the mating receptacle. You'll have to install a new connector to correct the problem.

There's a lot more information about how to install and operate scanners and external antennas for them in the author's *Two-Way Radios and Scanners For Dummies,* published by Wiley Publishing, Inc.

Weatherproofing an Exterior Connection

Stuff You Need to Know

Toolbox:
- Regular pliers

Materials:
- High-quality electrical tape (Scotch 33+ or 88 is good)
- Anti-oxidation compound (Penetrox, Oxygard, or similar formulation)

Time Needed:
Less than an hour

Even if you install a connector perfectly, after it gets exposed to the weather, water will infiltrate the smallest cracks and crannies. Soon the connection will become intermittent or worse! That's why this task shows you how to weatherproof a Type F or UHF connector installed at an antenna or at an connection panel.

Type F and UHF connectors are suited for outside use. BNC connectors are not — and should be avoided in exposed applications.

1. Put a very small amount of anti-oxidation compound on the threads of the receptacle connector (not on the cable plug connector). In addition to preventing water from getting in via capillary action, the compound also resists corrosion and makes the connector much easier to get off later. The compound is the smear of dark material at the top of the connector — it doesn't take much!

2. Put another small amount of compound on the center pin or center conductor of the cable's plug connector. Seat the plug in the receptacle and twist it back and forth a few times to smear the compound out around the pin. There should be no compound bridging the space between the center pin or conductor and the connector shell. If there is, remove the connector, wipe it clean, and start over.

3. Tighten the connector shell finger-tight and be sure the connectors are firmly seated. Use the pliers to tighten the shell another ⅛ to ¼ turn. This would be a good time to check and make sure the system is working properly before you waterproof the connection.

4. If the connection is horizontal and at the low point of the cable (such as an antenna cable coming down to a connection into your house), add a drip loop to the cable before you waterproof the connector. Any water running down the cable will drip off at the low point of the drip loop instead of being channeled directly into the connector!

5. If the connector has a big difference in diameters (50% or more) between the connector and the crimp or cable, This could make it difficult to get a smooth wrap, free wrinkles or gaps. Build up the smaller diameter by wrapping several times with tape. This will help make a smooth wrap.

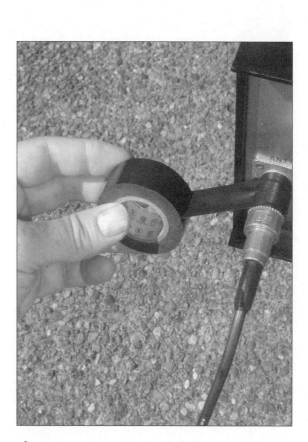

6. Begin where the threaded shell of the connector is attached to the receptacle. Start the wrap as far forward on the connector as possible. Wrap the electrical tape twice around this point and begin wrapping toward the rear of the connector with at least 50 percent overlap at each turn. This may be difficult in close quarters, so use a screwdriver or pencil to ease the tape into place. Use your fingers to press the tape into place.

7. Continue wrapping along the connector until you reach a point an inch or two behind the connector. At changes in connector diameter, give the tape plenty of overlap and press it into place with your fingers (on the non-sticky side only!) to minimize the number and size of wrinkles and folds.

8. At the end of the wrap, reverse course and wrap all the way back to the point at which you begun in Step 6. If the wrap seems wrinkled or it was difficult to seat the tape smoothly all the way around the connector, cut a short piece of tape and wrap one or two more times around the connector where the wrap ends.

It's not necessary (or even desirable) to pull the tape taut as you wrap! Good quality tape has plenty of adhesive and will seal just fine with little tension. Pulling on the tape as you wrap will cause it to *creep* as the plastic exerts tension on the adhesive over time. Just get the tape in place and snug.

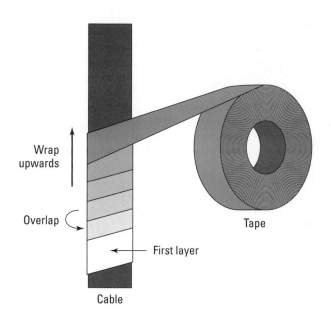

Wrap
upwards

Overlap

First layer

Tape

Cable

9. If the connection is vertically oriented, make your final wrap from the bottom up. As with shingles on a roof, the overlap will shed water down the cable and away from the connector. The opposite (wrapping top to bottom) captures water instead!

10. When you have finished the wrap, cut the tape end and press it firmly into place.

Don't pull on the tape until it breaks as a way of ending the wrap; remember the earlier caution about creeping tape. At the end of a wrap, creep leads to flagging where the tape's end eventually comes loose and waves around in the breeze. (Oh, say, can you see . . . ?)

Chapter 9

Mastering Power

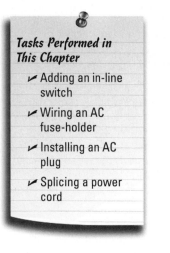

Tasks Performed in This Chapter

- ✔ Adding an in-line switch
- ✔ Wiring an AC fuse-holder
- ✔ Installing an AC plug
- ✔ Splicing a power cord

L ots of electronic stuff is powered from good old wall-socket AC voltage. That means lots of cords and connectors that can break or wear out or that have to be installed in new projects. This chapter shows you how to perform three common tasks for repair, maintenance, or new projects.

The goal of this first section is to teach you the rules of the road for AC wiring. It's not meant to frighten you but to give you the information to do the job safely and with confidence.

Note: This book does not cover household AC wiring, such as running cable for outlets and circuit breakers. If you have questions about performing this type of wiring, consult an electrician or other reliable electrical reference.

It's important to do AC wiring properly because AC power can supply a *whole lot* of energy. Faulty AC wiring — in the wall or in equipment — is a leading cause of fires. *Do not cut corners on any wiring connected to the AC line!* DO take your time and do the job right. It rarely takes much longer to use proper techniques, and whatever additional expense is incurred is cheap compared to the potential damage.

While you can use just about any old wire and components for the low-voltage DC part of electronics, the same is *not* true of wiring or components connected to the AC line. The insulation of wire used for AC wiring should be at least 600V. Voltage ratings for components connected to 120V circuits should be 180V or higher. Electronic components of all sorts should be *line-rated* and carry an Underwriters Laboratories (UL) or other safety marking. If you're in doubt, check with the product vendor.

Be Safe, Not Sorry!

Whenever working with or around AC voltage, remember its significant hazards of shock and burns. Most electronic projects use low DC voltages that pose little hazard (except to the pocketbook). It's easy to be lulled into complacency! Treat every wire as if it were connected right to the wall socket. Never assume that the power is off — confirm it with a multimeter. The author has avoided shocks several times by first making a simple test. Remember that electricity moves much faster than you do. It's not possible to make a swipe test with a finger to see if a wire is hot and *not* run the risk of a full-blown shock.

Try not to work on equipment with exposed connections to AC power. If you must (for example, during troubleshooting or repair), here are a few critical rules to follow:

✔ Follow the "one hand in the pocket" rule: When working on energized equipment, keep one hand in your pocket when making tests or adjustments.

✔ Never work in bare feet — wear shoes with insulating soles. The goal of both rules is to keep shocks from causing current to flow through your body. Currents through the body have the greatest potential to disrupt your heartbeat.

✔ If you're working on energized equipment, make sure someone else is around to turn off the power if you're shocked.

The Web site (`www.repair-home.com/Electrical_Systems`) has more information about basic electrical safety practices. Basic first-aid (`www.mayoclinic.com/health/FirstAidIndex/FirstAidIndex`) and CPR training (`www.mayoclinic.com/health/first-aid-cpr/FA00061`) are also good ideas for any electronic-er.

When you're working with AC line cords or power cords, make an extra effort to be sure you're following the right color codes and wiring conventions. To do so, you'll need to know what they are! First, some definitions:

AC wiring in the home consists of three wires: hot, neutral, and a safety ground (or just "ground").

✔ **Hot:** The wire that supplies current from the power source.

✔ **Neutral:** The wire through which current returns to the power source.

✔ **Safety ground:** The wire connected to the Earth. Neutral and safety ground are connected together at the main circuit breaker box. The metal chassis (a framework or enclosure) of equipment or appliances should be connected to the safety ground wire.

A color code is used to allow power wiring to be done correctly without having to use test equipment. By following the AC-wiring color code strictly and without exception, your wiring will be safe and correct. The color code for 120V household wiring is as follows:

✔ **Hot:** Black (terminals on plugs and sockets are gold or brass-colored).

✔ **Neutral:** White (terminals on plugs and sockets are silver or white).

✔ **Ground:** Bare or green (terminals on plugs and sockets are green).

(For color codes on 240V (and higher-voltage) circuits and cords, consult a book about household AC wiring.)

2-wire power cords also have a hot-neutral wiring standard. You may have noticed that on 2-prong plugs, one prong (or blade) is wider than the other so that the plug can only be inserted into the socket in one orientation. These are called *polarized* plugs; the wider prong is connected to neutral. In addition, on 2-conductor wire (or *zip cord*), the insulation on one wire is smooth (this is the hot wire) and the other ribbed (this is the neutral). Zip cord gets its name because the two wires can be easily pulled apart like a zipper.

Power and extension cords can be a major safety hazard if used improperly. Never use extension cords with non-polarized plugs or cords with broken-off ground pins. Never break off the ground pin on an AC power plug. If necessary, use a 3-wire-to-2-wire ground adapter (sometimes referred to as a *cheater*) — and if possible, connect that adapter's ground terminal to the outlet's faceplate mounting screw (which *should* be connected to ground).

Wire is also limited in the amount of current it can carry without overheating. If too much current flows in the wire, it will get hot enough to melt its insulation or start a fire. This is caused by the wire's resistance to current. Thicker wire has lower resistance and can carry higher currents. Be sure to use wire that's heavy enough to carry the current you're using, as shown in Table 9-1.

Wire thickness is measured as a *gauge* — a set of numbers with each number indicating a specific thickness. The U.S. standard for gauges is AWG (American Wire Gauge). Wire gets thicker as gauge gets smaller. For example, 8-gauge wire is thicker than 12-gauge.

Table 9-1	Current-capacity of common wire gauges	
Gauge (AWG)	*Current (Amps)*	
24	2.1	
22	5.0	
20	7.5	
18	10	
16	13	
14	17	
12	23	
10	33	

Table 9-1 contains information from Reference Data for Engineers: Radio, Electronics, Computer and Communications, 7th Ed.

Adding an AC In-Line Switch

You just completed a motorized corn-on-the-cob shucker and it works great! But you have to be able to turn it on and off, don't you? You can find the perfect solution in your local hardware store: an in-line switch that you install in the power cord between the motor and the wall socket! In this task, I walk you through the steps to make installation easy and have the final product look great.

There are actually two kinds of in-line switches for power cords. This task uses the big kind — about the size of an egg. It fits power cords up to those with three 12 gauge wires, can switch several amps of current, and has a sturdy switch that snaps on and off. The small kind is made for 2-wire power cords and only handles light loads, such as table lamps. If you learn how to install the big version, the small version will be easy. Just be sure to choose a switch that's rated to handle the current consumed by whatever it's controlling.

In AC power circuits, switches and fuses are *always* installed in the hot line, not the neutral. The hot wire is the source of power, so you must interrupt power in the hot wire when turning the device on and off. Switching or fusing the neutral leaves the hot wire connected to the equipment — a definite safety hazard.

1. Take the switch apart by loosening the screws. Place the screws, nuts (if any), and cover (the part of the shell without any switch parts in it) in a can or box for safekeeping until you're ready to reassemble the switch. Be careful not to lose the nuts as they often lodge in the cover or switch body. The exposed switch will probably have a little grease on some of its parts — don't wipe it off!

2. Read the instructions that come with the switch. The following instructions apply to most switches, but you might have to modify them a little for your particular model.

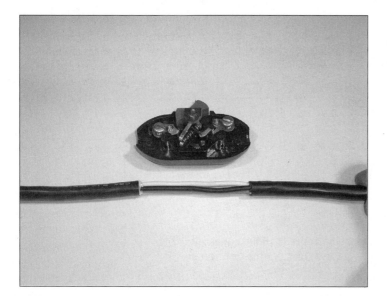

3. Find the point in the power cord at which you want to install the switch. Remove a length of the jacket equal to the length of the switch minus ¼". To remove the jacket without nicking the inner wire insulation or strands, use a knife to score the jacket until it parts when the power cable is bent at the cut. Make a shallow cut along the section of jacket to be removed and peel the jacket apart so that it can be removed. Untwist the cord so that the wires can be easily separated. Remove any string or fiber filler material from the exposed area.

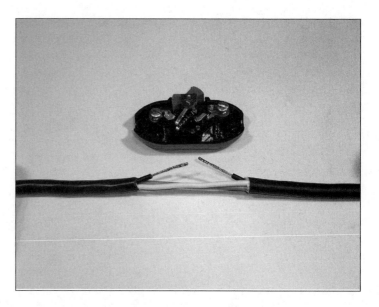

4. Cut the hot (black) wire in the middle of the exposed area. Strip the black wires so that only ¼" of insulation is left on the wire as it exits the jacket.

If you're installing a small switch on 2-wire zip cord, the hot wire is the one with the smooth (non-ribbed) insulation.

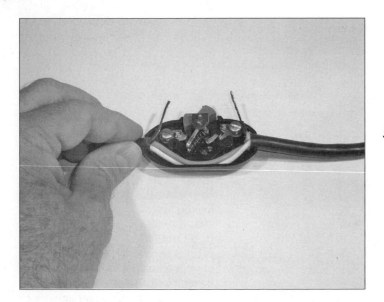

5. Lay the entire cord in the body of the switch so the neutral (white) and ground (green) wires are in the groove bypassing the switch.

6. Loosen the connection screws of the switch so there is plenty of room for the wire, but don't remove them entirely. If the screw comes out, reinsert it.

7. Wrap the stripped ends of the hot wires clockwise three-quarters of the way around the screws. Don't wrap the wire all the way around the screw — and definitely *not* more than once. The wrap should be clockwise so that as the screw is tightened, it pulls the wire around and under the screw instead of pushing it out.

8. Tighten the screws and trim the excess wire.

9. Look closely at the exposed switch body. Next to the holes on either end where the power cord enters the switch, a small rib just inside the hole acts as a strain relief by capturing the power cord jacket when the cover is attached. Push the cord in slightly at each end so the jacket is at least ⅛″ past the ribs.

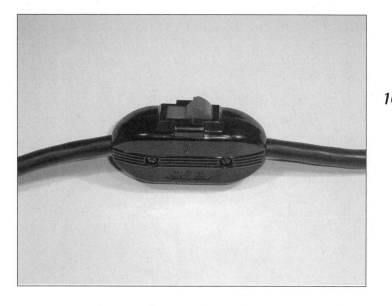

10. Reattach the switch cover and make sure the individual wires are not exposed at either end of the switch. If they are, remove the cover again and reseat the cord in the hole. Once the cover is attached, you can plug in the cord and try it out!

Wiring an AC Fuseholder

For any project that brings AC power directly into the equipment case (as opposed to power from an external supply), a fuse is a *must* — both for safety and to protect the equipment from damage in the event of a short circuit or component failure. Chapter 7 shows how to install an in-line fuse holder in a DC power lead — for AC circuits, however, it's better to use a chassis-mounted fuse holder for safety. In this simple task, you'll learn not only about attaching wires to terminals, but to think about safe ways of dealing with AC power.

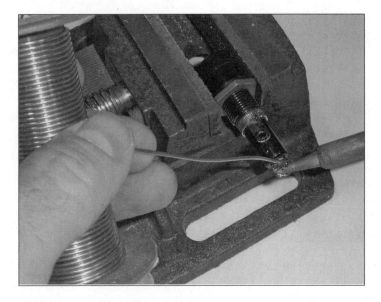

1. Turn on the soldering iron and when it's hot, *tin* (coat with solder as described in Chapter 2) both terminals of the fuse holder.

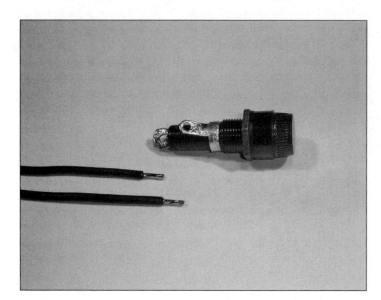

2. Strip the ends of the wire that will attach to the fuse holder and tin them.

3. Cut two pieces of heat-shrink tubing about 1″ long. The tubing should fit over the fuse holder terminals. Slip the tubing over the wires.

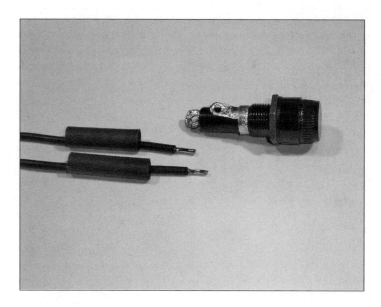

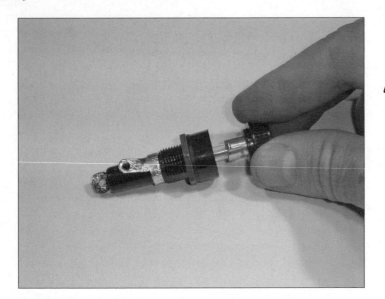

4. Look carefully at the fuse holder, noting that there is one terminal at the very back (at the left of the figure) and a forward terminal (at the right of the figure) just behind the threaded portion of the body. Place a fuse in the cap of the fuse holder and fully insert it. Now remove the fuse slowly, stopping about halfway out.

5. If you are installing the fuse holder in a piece of equipment, orient it with the terminals oriented so that it's easy to solder the wires to them. (A vise is used in the figure for the following step so you can see the usual orientation.)

6. Insert the tinned live wire into the rear terminal of the fuse holder and solder it.

One end of the fuse cartridge is in the cap, held in your fingers. The other end is halfway into the holder, in contact with the forward terminal — and your fingers are perilously close to the metal end of the fuse! If the hot wire is connected to the forward terminal, it is easy to get shocked when removing the fuse. The hot wire should *always* be connected to the rear terminal of the fuse holder. In other words, power should flow into the back of the fuse holder and out through the forward terminal so that if the fuse is removed while the equipment is energized, AC voltage is never present on the exposed cap of the fuse holder.

7. Slide the heat shrink tubing over the terminal and shrink. it onto the terminal. It is OK if the portion over the wire is loose as the goal is to insulate the exposed terminal.

8. Repeat Steps 6 and 7 for the remaining wire, and then install the fuse.

Installing an AC Plug

The simple task of replacing or adding an AC plug to a power cord is another seemingly-trivial task often performed poorly by home technicians. This task shows you techniques for working with AC power connections — and gives you some practice at using them. You'll also reinforce your knowledge of color codes.

Before beginning, make sure you've chosen the right size plug (and cord or cable) for the job. Most 3-prong AC sockets and plugs are not designed to carry more than 20 amps. If you're going to be supplying more power than that, you should use a high-current plug and socket. Plugs and sockets for ac power have a "NEMA" (National Electrical Manufacturers Association) rating number that shows what type they are and how much current and voltage they can safely carry. The rating is also stamped into or molded in the body of the plug or socket. For more information, the Wikipedia entry on NEMA connectors (http://en.wikipedia.org/wiki/NEMA_connector) will provide all the information you need.

Heat is the enemy of electrical connections. Run your vacuum cleaner for a while and then unplug the cord, feel the AC plug, and feel the socket: They're warm! The heat comes from electrical resistance in the connections of the wires to the plug and socket contacts and in the contacts themselves. The higher the current, the higher the heat given off. If overloaded, contacts can get hot enough to melt or char the insulation around them. The metal of the contacts can oxidize, further increasing resistance and heat. Overloaded contacts will quickly fail, creating a significant fire hazard. Use properly rated cable and plugs to avoid this very real hazard.

How do you stop zip cord from unzipping itself? Zip cord, used for 2-wire power cords and speaker cables (among other uses), is made so the conductors will separate easily and cleanly. Sometimes they separate a bit too easily. To hold the wires together, learn the Western Union knot shown in Figure 9-1. With about 2″ of the cord unzipped, loop one so the wire comes back across the cord on the side facing you as shown in Step 1. Loop the other wire around the free end of the first, behind the cord, and back through the loop of the first wire as in Step 2. Each wire is then captured in the loop formed by the other wire. Gently tighten — it's not necessary or desirable to make the knot really tight — and your unzipping worries are over!

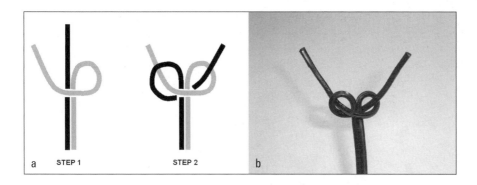

a   STEP 1   STEP 2   b

1. Separate the shell of the plug from the contact assembly by loosening the screws in the front of the plug body next to the prongs. (Your plug may vary from the one shown in the figure.)

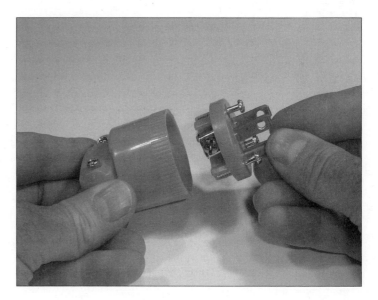

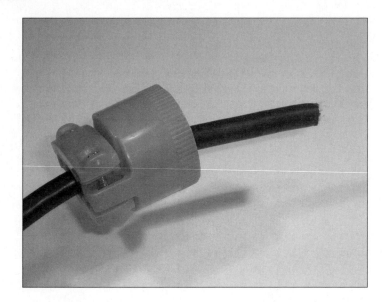

2. Slide the shell onto the power cord. You may have to loosen the screws on the strain-relief clamp that holds the cord. It's not necessary to completely disassemble the clamp. (Some plugs have metal clamps, or the clamp may be built into the shell of the plug.)

3. Remove 1″ to 1½″ of the power-cord jacket. When the plug is completely assembled, the strain-relief clamp should make contact only with the jacket and none of the inner wires should be exposed. Check the amount of jacket to be removed by sliding the plug parts together and taking a look. (Your plug may have come with instructions that specify how much jacket to remove.) Avoid nicking the inner wires or their insulation by scoring the jacket and flexing the cord until the jacket breaks away cleanly. Trim off and discard any string or fiber filler.

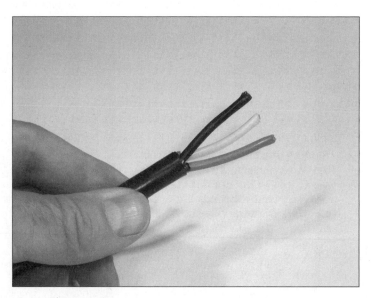

4. Strip ½″ of the insulation from each wire. Twist the strands lightly to keep them together, following the manufacturer's direction of twist. Inspect the wires to be sure that they are shiny and not corroded or oxidized. If they are, cut off 6″ of cable and go back to Step 3. The wires must be clean and bright to minimize resistance at the connection.

5. Loosen the terminal screws on the plug's contact assembly. Do not remove the screws, just loosen them enough so the stripped wires can be inserted.

6. Arrange the wires so that they line up with the correct terminals; hot (gold), neutral (silver), and safety ground (green). If your plug has screw terminals, one terminal at a time, wrap the stripped end of the wire clockwise around the screw just ¾ of a turn. If instead your plug has clamps that capture the wire, insert the stripped portion of the wire into the clamp (no insulation should be in the clamp) and tighten.

Don't wrap the wire all the way around the screw — and definitely *not* more than once. The wrap should be clockwise so that as the screw is tightened, it pulls the wire around and under the screw instead of pushing it out.

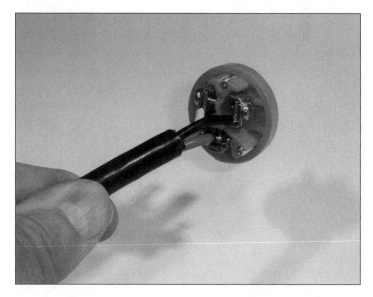

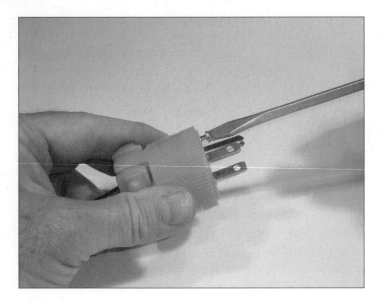

7. Slide the shell of the plug back onto the contact assembly and attach it.

8. Tighten the strain relief clamp onto the jacket of the cord. If you cut the jacket a little bit long (less than ¼"), you can push the cord into the plug and then tighten the clamp. More than ¼" and you'll need to disassemble the plug, cut the wires a bit shorter, and try again. Don't tighten the clamp so much that it bites into the jacket. It should be tight enough that the cord doesn't move inside the connector when you push or pull on the cord.

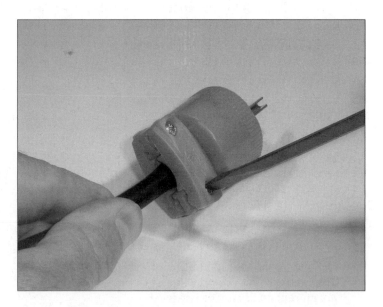

Splicing a Power Cord

Power cords get damaged from all sorts of things; over-enthusiastic weed-whacking or hedge-trimming, forgetting that the car's battery warmer was still plugged in, even being chewed up by dogs and rodents! New cords aren't cheap as any hardware store patron knows. This task shows you how to make that cord almost as good as new for the cost of a little time and effort.

This task shows how to repair a 3-wire power cord but the same techniques apply to their smaller 2-wire power cord cousins.

1. Trim the separated ends of the cord and remove any frayed, wet, or oxidized wire.

2. Cut a 6" piece of heat-shrink tubing with a diameter 50% greater than the cord's jacket.

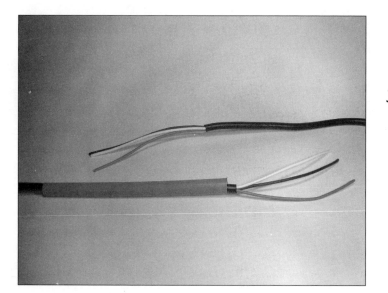

3. Trim 4" of jacket from each piece of cord. Avoid nicking the inner wires or their insulation by scoring the jacket and flexing the cord until the jacket breaks away cleanly. Untwist and straighten each wire. Remove any string or filler material. Slide the heat shrink tubing over either piece of cord.

4. To separate the splices of each individual wire, perform the splice at a different point on each pair of wires. On the left-hand piece of cord, cut the wires as follows: hot — 2" from the jacket, neutral — 3" from the jacket, ground — do not cut.

   On the right-hand piece of cord, cut the wires as follows:

   • Hot — do not cut

   • Neutral — 3" from jacket

   • Ground — 2" from jacket

   Strip each wire 1". You can see in the figure that each splice will be 1" apart, keeping plenty of insulation between the bare wires.

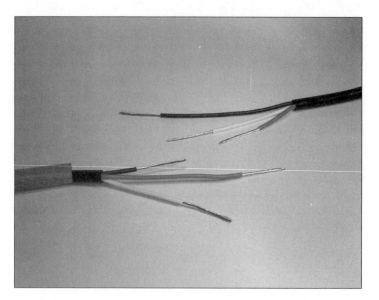

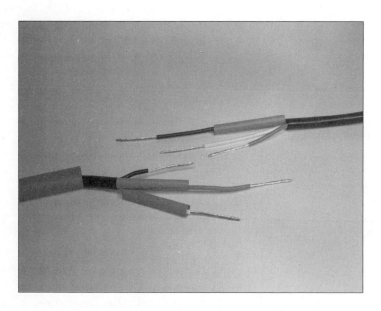

5. Cut three 1½" pieces of ¼" diameter heat shrink tubing. Slip one piece over each wire.

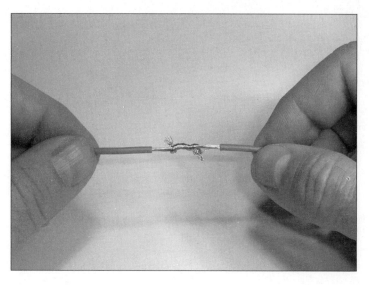

6. Starting with the neutral wire (it will be the easiest since both pieces are longer), cross the stripped wires at their midpoint and twist each around the other in opposite directions. This is a "Western Union" splice, originally invented for repairing telegraph lines in the mid-1800s! Continue twisting until the wires are snugly wrapped around each other.

7. Solder the wires together, center the heat-shrink tubing over the splice and shrink it.

8. Repeat Steps 6 and 7 for the hot and ground wires.

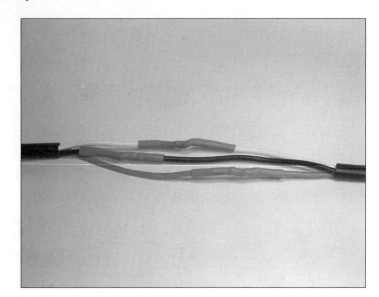

9. Center the large piece of heat-shrink tubing over the entire splice and shrink it.

10. Wrap the splice with good-quality electrical tape (Scotch 33+ and Scotch 88 are excellent choices) twice, once in each direction, over-lapping each successive wrap at least 50%. At the end of the wrap, cut the tape and press it into place.

Do not pull the tape apart; that causes the top layer to begin to separate. Eventually pulled-apart tape comes completely loose (*flagging*).

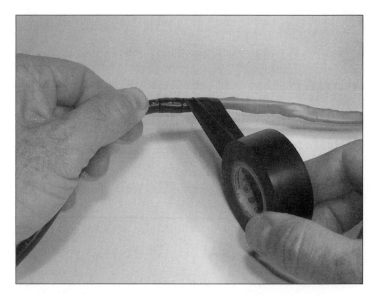

Chapter 10

Audio and Sensitive Connections

Cables and connectors that carry low-level audio and instrumentation signals need special care, even if they look rather ordinary. The small signals that they carry are easily contaminated with noises and buzzes if the cable or connectors are compromised in some way. In this chapter, you'll work with three common audio connectors and make a handy temperature sensor as you get some practice and tips.

The XLR: A Real Pro Connector

What is unique about microphone cables? Well, for openers, they get more abuse and handling than any other type of cable you're likely to encounter! Think about what performers and speakers do to microphones; they yank them in and out of the stand, drop them, drag them, twirl them around by the cable, then wad the whole thing up and stick it in a box or bag, expecting it to all work perfectly next time! The result is often a dead cable (no audio), humming, mixed in with or replacing the audio, or intermittent or scratchy audio.

The XLR connectors seen in Figure 10-1 are professional quality. If you ever get to be a roadie for a band, you'll see plenty of these connectors on thick, black cables that plug into amplifiers and audio mixing boards. (You'll also fix a lot of them.) An XLR connector has three contacts and a heavy metal (no pun intended) shell. This combination makes it tough and electrically secure.

The ends of cables generally have female XLR connectors while equipment and wall plate connectors are male. (You will come across exceptions.) The shell is not electrically connected to any of the audio pins — it's for mechanical protection only. Having all of the electrical signals inside the connector means that it doesn't matter what the shell bumps into, such as another audio signal or a power cable that might cause noise or buzzing.

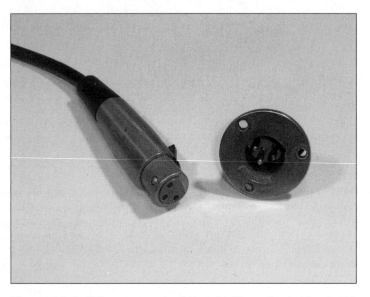

Figure 10-1: XLR connectors are the standard for professional and public-address system microphone and audio wiring.

Standard connection conventions

The standard connection conventions are also shown in Figure 10-1. You should follow this convention for all your cables. If you discover a particular piece of equipment with another wiring convention, you can either change the equipment or make a short *adapter cable* that moves the wires to their required new homes. By keeping all your cables consistent, it's easy to hook up a portable audio system without worrying about wiring conventions all the time.

Balanced/unbalanced wiring: Audio wiring is called *balanced,* when neither one of the twisted-pair conductors is connected to ground; instead the wires are twisted together to reject electrical noise and interference. They are surrounded by a shield braid or foil for further protection from noise. *Unbalanced* wiring uses ground as one of the conductors that makes up the complete circuit — often the outer shield. Coaxial cable is an example of unbalanced wiring.

Do not directly connect either of the signal conductors in a balanced system to ground. This unbalances the formerly balanced signals and usually results in hum or distortion.

Where do hum and buzz come from?

The weak hum you hear in many amplifier outputs is caused by 60-Hertz magnetic fields from power lines and cables. Audio cables act like terrific antennas; they pick up these fields from the transformers in power supplies and equipment, lighting, motors, and any kind of electrical equipment. You can minimize this type of hum by routing microphone and audio cables away from power circuits and equipment.

Louder buzzes and hums are often the result of *ground loops* or broken shields in the cable. Start by checking the grounding and shielded connectors and cables. Loud hum is often the result of *ground loops,* in which unequal grounds between pieces of equipment (often plugged into different power outlets or circuits) allow 60-Hertz current to flow in the shields. It can also result from broken cable shields or ground wires. If rerouting your cable doesn't fix the problem, give that connector a gentle shake or tug while it's plugged in; see whether you can hear crackles or pops that result from short-term bad connections.

Plugging In to Phono Plugs

Also known as *RCA connectors,* these very common connectors shown in Figure 10-2 are found on patch cables for audio and video. The back of a DVD player or stereo receiver has quite a number of phono jacks for inputs and outputs, grouped together by function (such as CD IN, VIDEO 1, and so on). Each carries a single (or *mono*) channel of audio or a single channel of video. This task shows you how to install a new phono plug on an audio or video cable that plugs into one of these jacks.

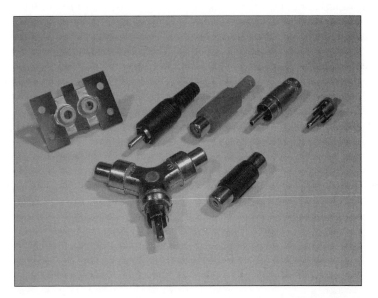

Figure 10-2: Phono or RCA plugs and jacks are widely used for both audio and video connections.

Why is it called a "phono" plug?

Phono is short for *phonograph*. After the early days of radio, the home receiver became more and more powerful, sophisticated, and fancy — a real piece of furniture. It was natural to want to take advantage of its high-fidelity (for the time) audio amplifier and speaker to play records, which were originally played on standalone Victrolas. Electrical signals from the tone arm pickups had to be connected to the radio's audio amplifiers. The solution to the connection problem, pioneered by the Radio Corporation of America (RCA), was a simple cylindrical connector (there was no stereo back then) with a center pin and a shell. These connectors were labeled "Phono" and became very closely identified with RCA, the Microsoft of its day in size and marketing prowess.

Patch cables for audio and video are so inexpensive that unless you're making a special custom cable or need to repair a cable quickly, it's much more cost-effective simply to buy new cables. The next time you're in an electronics store, buy a few spares so you're never caught without spare or replacement cables.

It's natural to want longer patch cables, but beware: Cables longer than about 15′ place loads on equipment inputs and outputs that can reduce high-frequency response, making audio sound muffled or video to be blurry and jumpy. For longer separations of audio connections, use low-impedance line-level inputs and output. For video, use coaxial cable and 75-ohm inputs and outputs.

Phone plugs (as shown in Figue 10-3) are descendents of the connectors used at telephone switchboards (see "Installing a Telephone Plug" in Chapter 7) and are still common in audio equipment and circuits today. Along with the ¼″ model, there are also miniature 3.5 mm and subminiature 2 mm versions. All three are available in stereo (3-wire) and mono (2-wire) versions. The common standard is for the connector's tip contact to carry the left audio channel and the ring contact the right audio channel. The sleeve is a common ground for both channels.

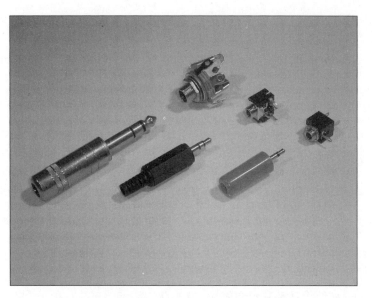

Figure 10-3: Phone plugs come in three sizes: ¼", miniature ⅛" (or 3.5 mm), and subminiature 2 mm. They are used for audio and control connections. Their mating jacks are shown to the right of each plug.

Taking a Temperature Electronically

Most of the connectors in this chapter are for cables that carry audio signals. Another type of sensitive signal carried by cables is a DC voltage or current that represents the value of some measurement. These signals also need protection because noise or disruption degrades the accuracy of the measurement.

The temperature *sensor* in the task "Constructing a Temperature Sensor" later in this chapter (an LM34C) is an IC built into a small plastic package (known as a TO-92) of the type that usually holds a single transistor. The internal circuit is simple, and the small package helps the sensor respond quickly to changes in temperature. Download the data sheet for the LM34C at www.national.com/ds/LM/LM34.pdf#page=1 to use as a reference.

Sensor: A device that measures the value of some physical parameter (temperature, flow, level, brightness, and so on) and outputs the measurement as an electrical signal, usually voltage.

Temperature Sensor Wiring Diagram

The circuit for the temperature sensor in Figure 10-4 is very simple. All of the internal circuitry (see the Block Diagram at the end of the downloaded LM34C data sheet) is contained inside the TO-92 package. The LM34 only has three connections; power, output, and ground. To output a voltage representing temperatures from +5 to +300° F it can be powered by voltages as high as 35V DC. (Special circuits are shown in the data sheet to allow measurement of temperatures below zero degrees.) You can read the output voltage of the circuit with a multimeter and interpret it as a temperature: 1.00V = 100° F.

The only non-obvious part of the circuit is the 2.2 kΩ resistor at the output connection. This is a *decoupling resistor* that isolates the sensor from the small capacitance created by the cable connecting the sensor to the voltmeter. (See "Capacitive Loads" on page 6 of the IC data sheet for more information.) The IC is very small; its output circuit can't charge up that small capacitance and still keep the voltage steady. Without the resistor, the output becomes unstable, jumping back and forth as the sensor output circuit tries to maintain a steady output voltage but can't quite keep up. This creates a *galloping* effect as the output voltage swings back and forth a volt or two. The resistor limits the amount of current the output can pump into the cable — and that calms things down so the output can maintain a steady voltage at the cost of a very small voltage drop caused by the resistor. (It won't hurt the sensor to be operated without the resistor, so feel free to try it and see what happens!)

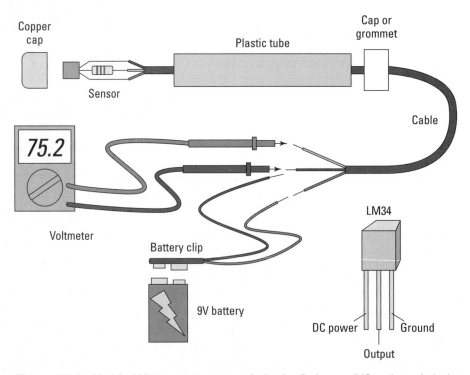

Figure 10-4: All of the LM34 temperature sensor's circuitry fits in a small IC package. A single resistor and battery create a voltage representing temperature.

Installing a Microphone Connector

In this task, I assume that you've been given a bad microphone cable with a standard XLR connector. You'll disassemble, check, repair, and reassemble the cable and connector. In the process, you'll become familiar with this reliable, sturdy connector and the standard wiring conventions for it.

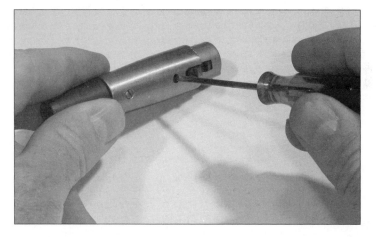

1. Disassemble the old connector by screwing *in* the small set screw located near the front of the connector. Turn the screw as if you were loosening it (counter-clockwise) and the screw will move farther into the central plastic body of the connector. If your connector has cable clamp setscrews (usually two) at the rear of the connector shell, loosen but do not remove them.

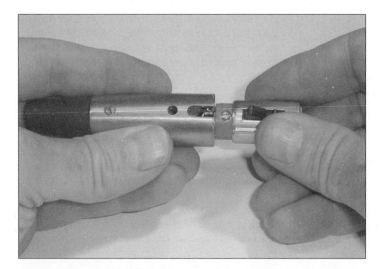

2. When the screw head has withdrawn past the inside surface of the metal shell (you can see this by looking into the set screw hole) gently pull the back portion of the connector out of the front portion while rocking it back and forth. This will expose the contact terminal block of the connector. Push the back part of the shell farther along the cable until the connections are completely exposed. There may be a plastic insulating sleeve around the terminals. Push that back along with the shell. Keep the shell on the cable.

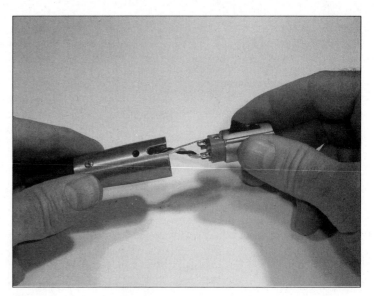

3. When you are repairing a cable, this is the point at which you can inspect the connections between the wires and shield and the terminals. A bad cable will usually have one or more connections completely broken, have only one or two strands of wire left connected, or have a short circuit between two terminals from loose strands of wire.

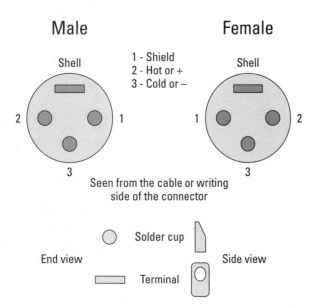

Male

Shell

1 - Shield
2 - Hot or +
3 - Cold or −

2   1

3

Female

Shell

1   2

3

Seen from the cable or writing
side of the connector

Solder cup

End view   Side view

Terminal

4. On a piece of paper, write down which color of wire is soldered to which terminal. Then unsolder the wires from the terminals; set the block with the terminals aside if you're going to reuse the connector. Cut off the end of the cable completely, about an inch back from where the jacket ended. This also removes any worn area of the shield. If you're going to use a new connector, replace the rear shell, strain relief, and any insulating sleeve with those from the new connector.

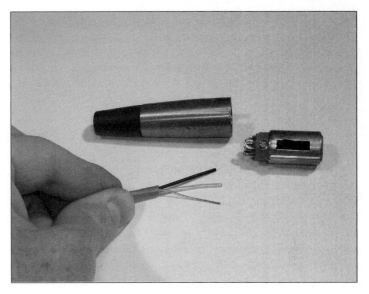

5. Remove 1″ to 1 ½″ of jacket, being careful not to nick the braid or foil. Trim away any string or fiber filler in the cable. If the shield is foil, peel back and remove the foil while leaving the shield wire intact. If the shield is braid, twist the strands together to make one fat wire and tin the end to keep it together.

6. Strip ¼″ off each wire and tin.

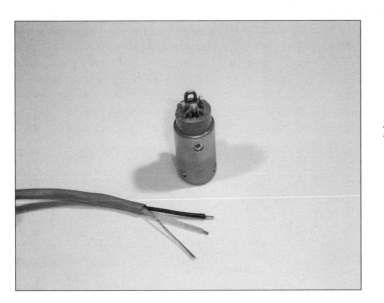

7. If you are reusing a terminal block, remove all excess solder and wire bits from the terminals. For a new connector tin the terminals.

8. Connect each wire, including the shield wire or shield, to the terminals as noted in Step 4. If present, slide the insulating sleeve back over the terminals. If your connector had the shell terminal (the large, un-numbered terminal) jumpered to pin 1, make that connection by pulling the shield wire or braid through the shell terminal to pin 1. Solder the shell terminal first, then pin 1.

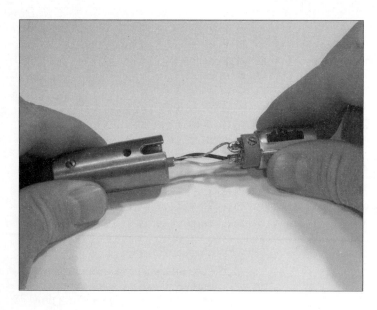

9. Pull the cable through the back of the shell to seat the terminal block into the shell. You may have to turn the shell in order to fit the cable in with the right orientation.

10. Slide the front part of the connector back on and align the access hole in it with the screw hole for the set screw.

11. Turn the set screw to the left so that it backs out of the terminal block into the shell, but no farther than its surface being flush with outer surface of the shell. If there are any clamps at the back of the connector, tighten them now.

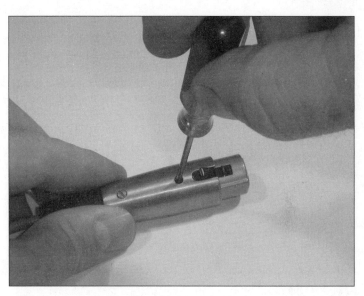

12. Check continuity from end to end of each contact to be sure the cable is wired properly: Both ends should be exactly the same, and there should be no short circuits between contacts.

If the new cable has the same problem as the old cable, you probably have a bad connector on the audio equipment. Swap in a different cable and see if it acts any differently. If not, the equipment connector is probably bad. Otherwise open up your cable and check again.

Installing a Phono Plug

Phono plugs and cables are widely used for audio, power, control signals, and even low-power transmitting equipment. It's quite easy to install and repair them — a good thing since they are not particularly sturdy. In this task you'll install a phono plug on a shielded audio cable, typical of most wiring that uses phono connectors.

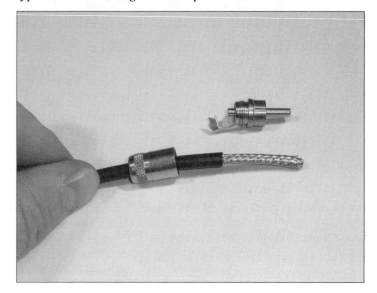

1. Trim the end of the cable and strip away an amount of the cable jacket equal to the distance from the tip of the connector to the forward edge of the strain-relief wings. If the phono connector has a removable shell, slide it on to the cable now.

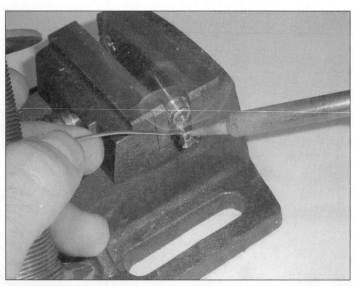

2. Tin a short section of the phono connector body halfway between the strain relief wings and the connector body. This is where the cable shield will be connected.

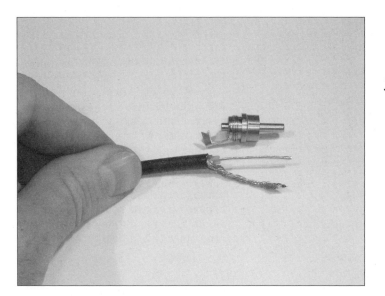

3. If the cable has a braided shield, comb out the strands. Twist the shield wires together to make a single wire. Tin one end to keep it together. If the cable has a foil shield, remove the foil along with the jacket and use only the single-shield wire. Strip and tin the cable's center wire, leaving enough insulation to prevent short circuits to the shield.

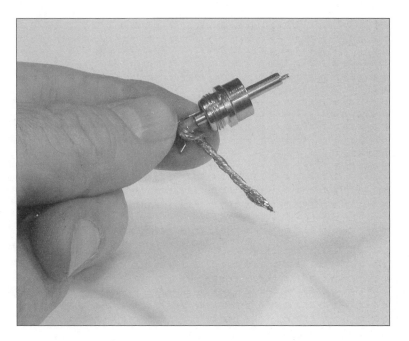

4. Insert the center wire into the phono connector so the wire just barely sticks out the front of the center pin and the end of the jacket is just forward of the strain-relief wings. Make sure there are no loose strands of wire that might cause a short circuit to the connector shell.

5. Solder the center conductor to the tip of the metal pin and trim away any of the wire sticking out past the pin. It helps to have the phone connector tilted slightly downwards so that solder will stay close to the connector's tip. Solder on the outside of the pin may increase the pin's diameter, making it hard to insert into the socket. Carefully shave off excess solder with the knife. Clean any solder flux off of the pin with a knife or brush.

6. Bend the shield wire so it lies on the tinned area of the connector. Trim away any excess wire. Solder it to the connector.

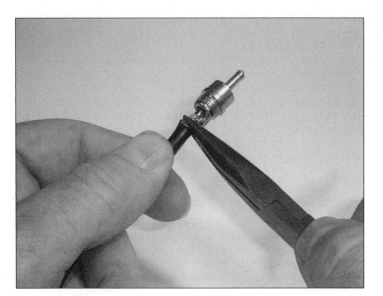

7. Wait a minute or so for the connector to cool. Use needle-nose pliers to capture the jacket of the cable in the strain-relief wings. Do not force the wings into the jacket; just clamp the cable. Trim any loose strands of braid wire.

8. Reattach the connector shell and use a voltmeter to measure resistance of the cable between the center pins and between the metal bodies — both should show short circuits. Check between the center pin and the shield — that should be an open circuit.

If your connector shows a short circuit between the center pin and connector body, the most likely cause is a stray strand from the center wire. You'll have to unsolder the shield wire, unsolder the center pin to remove the cable, and check for stray strands or bits of solder. Leaving solder flux on the center pin can cause an erratic or non-connection when the connector is inserted into a mating socket. If the shield connection is erratic, the usual cause is a poor solder connection of the shield wire.

Creating a Stereo Patch Cable

It's frequently useful to connect the headphone output of an audio source such as a radio or music player to the line-level stereo input of a computer sound card or recorder. Both of these devices usually have a 3.5mm miniature stereo phone plug as their audio connector. In this task you'll learn how to make the appropriate cable to make the connection. You can use this skill to make all kinds of audio cables.

When you're making a stereo audio patch cable, you must use shielded twisted-pair cable that is intended for stereo audio signals. Audio cable has two separate wires surrounded by a metal shield, which is itself surrounded by the cable's outer jacket. (You can also use dual-pair cable that has separate wire/shield pairs for each channel.) Do not use unshielded cable because it exposes the signal wires to interfering signals and other noise. Cable with a braided or twisted shield made of many fine wires is preferable for patch cables because of its flexibility. An alternative style has a shield made of foil, with a separate wire in contact with the foil. The wire can be soldered to connectors. Either will do, although the foil-shielded cable is stiffer.

1. Start by unscrewing the connector's plastic or metal *shell* (also known as a *backshell*) to disassemble the connectors. Place the connector in your vise and tin the tip and ring contacts with solder. (See Part I for information on soldering.) Tin the sleeve contact just in front of the cable strain relief wings.

2. Use the wire strippers or knife to remove 1/2" of the cable jacket. Try not to nick the thin shield wires. Remove the shield foil, if present. If the shield is braid or fine wires, Twist the shield wires together to make a single twisted wire.

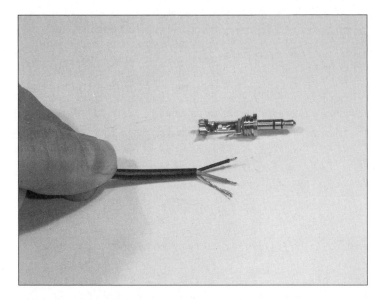

3. Strip ⅛" from each audio wire. Tin the stripped end of both wires, but leave the shield or shield wire untinned.

4. Cut two 1" pieces of heat-shrink tubing that will fit the cable jacket snugly when shrunk. (The tubing's unshrunk inside diameter should be about 50–100% larger than the jacket. Slip one piece over the end of the cable.

5. Slip the shell of the connector over the end of the cable, oriented so it will screw back onto the connector. Double-check to be sure you have it right; an error will require removing the connector to turn the shell around.

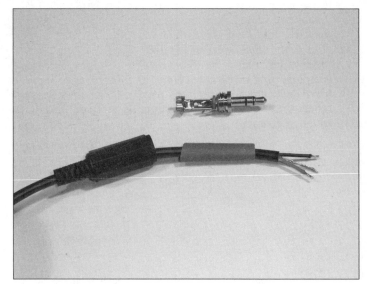

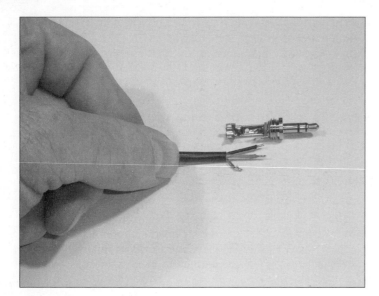

6. Place the cable in position and determine where the shield wire will touch the tinned area of the sleeve contact. Tin that portion of the shield wire and clip off the rest.

7. Prepare to solder the wires and shield to the connector in this order: shield, ring, tip. It will not be necessary to apply a lot of heat because both the wire and connector contact are already tinned. Place the tinned shield wire against the tinned area of the sleeve contact while holding (gently!) the insulation of the wire with the needle-nose pliers. Apply the soldering iron's tip; when the tinned area of both wire and connector have melted, withdraw the iron, holding the shield wire steady until the solder solidifies. Now hold the end of the ring wire to the connector's ring contact and do the same; repeat this procedure with the tip wire and contact.

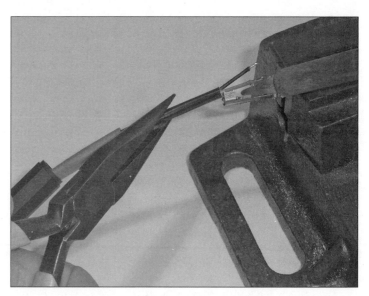

8. Wait for the connector to cool, then slide the heat-shrink up the jacket until it hits the connection of the shield wire and sleeve. Shrink the tubing onto the jacket.

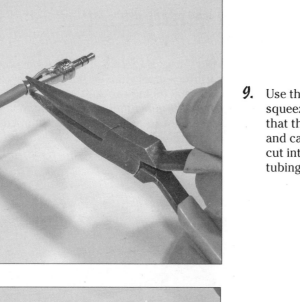

9. Use the needle-nose pliers to squeeze the strain-relief wings so that they capture the heat shrink and cable jacket. They should not cut into either the heat-shrink tubing or the jacket.

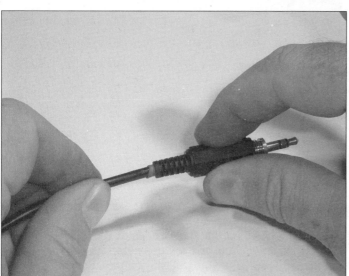

10. Screw on the shell. Repeat for the other end of the cable, if necessary.

11. Test the cable with a continuity checker or multimeter with resistance scale. Check tip-to-tip, ring-to-ring, and sleeve-to-sleeve — all should show low resistance of less than 1 ohm. Then check tip-to-ring, and sleeve-to-ring — all should show an open circuit (keep your fingers off the electrical contacts!).

On your first attempt, you may find it difficult to handle the small wires with the pliers without melting the insulation from the heat of the solder. This is a skill that takes a little practice! Learn to use the minimum amount of heat and the shortest soldering duration necessary so the insulation of the wires doesn't get so hot. If your resistance checks show short circuits between contacts on the same connector, try using less solder — a little goes a long way!

Constructing a Temperature Sensor

We'll build the sensor into a small metal tube, power it with a 9V battery, and read the temperature with a voltmeter. The sensor itself is mounted in a metal tubing cap so it will quickly reach the same temperature as the probe's surroundings. A plastic tube protects the assembly and allows the probe to be positioned without conducting heat toward (or away from) the sensitive sensor.

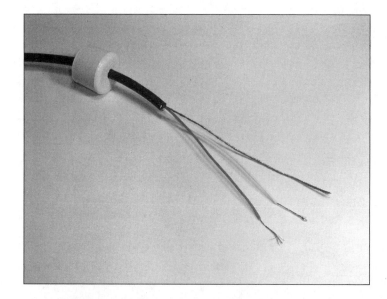

1. Prepare the probe's output connections by removing 3" of the jacket from the cable and any unused wires. Strip the three wires about ¾". (If you are using cable with a shield, use the shield as the ground connection by removing any foil or twisting the shield wires together and tinning the end.) Tin the stripped wire you will be using as the signal output wire. Leave the other wires untinned. Slide the grommet on to the cable. (If you are using a plastic cap, drill a small hole in it and slide it on to the cable.)

2. Twist the bare end of the 9V battery clip's black wire together with the wire you'll be using as ground for the LM34. Solder them together.

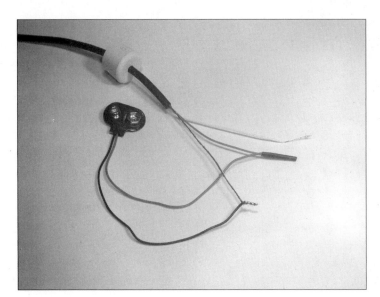

3. Twist the bare end of the 9V battery clip's red wire together with the wire you'll be using as power for the LM34. Solder them together, then place a short piece of ³⁄₁₆" heat-shrink tubing over the connection and shrink it on to the wires.

4. On the other end of the cable, remove 2" of the jacket and trim away any unused wires.

5. By now the epoxy should be cured so check to be sure that the LM34 is securely mounted in the cap. If so, cut the center lead of the LM34, leaving ½" extending from the body of the sensor. Lightly tin that lead.

6. Cut off one lead of the 2.2 kΩ resistor ¼" from the body of the resistor and lightly tin that lead.

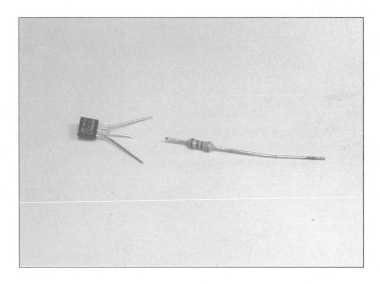

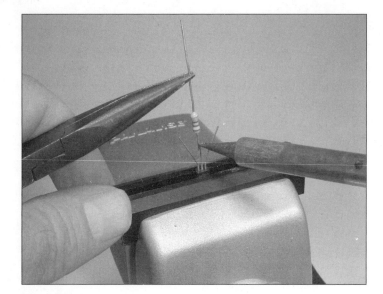

7. Hold the resistor with needle-nose pliers with the short tinned lead next to the LM34's tinned center lead. Melt a small ball of solder on the tip of the soldering iron and touch it to both leads. The solder will flow onto both leads. Remove the iron and hold the leads steady for a few seconds until the solder cools. The soldering step should only take a couple of seconds — don't hold the iron on the leads very long.

8. Trim the remaining lead of the resistor to ½" long and lightly tin it. Also tin the two remaining leads of the LM34.

9. Trim the leads of the cable to match the lead lengths of the LM34, remembering which lead will connect to which LM34 pin. (Review the IC connection drawing in Figure 10-4 earlier in this chapter to be sure.) Strip each lead about ¼" and tin.

10. Cut three 1" pieces of ⅛" heat-shrink tubing and slide them on to the wires protruding from the cable jacket.

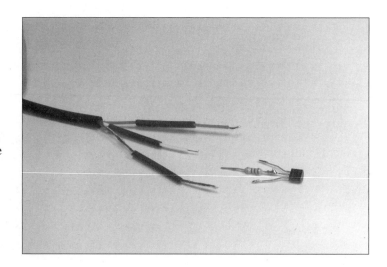

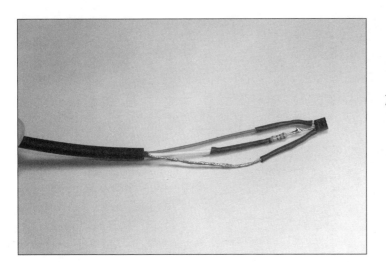

11. Solder each wire onto the appropriate lead of the LM34. After each wire is soldered, slide the heat-shrink tubing over the connections. Slide the tubing on the power and ground leads all the way up to the LM34 body and shrink the tubing.

12. Test the sensor before final assembly by connecting a 9V battery to the battery clip. Use a voltmeter to measure the output voltage between the ground and output wires. If room temperature is about 75° F, the output voltage should be a steady 0.750V DC plus or minus 30 mV (0.030V). If the sensor is working, disconnect the battery and voltmeter. If not, trace your wiring and be sure you have the proper connections and that none have come loose at the LM34.

13. Clean the inside of the copper tubing cap and place it open side up on your workbench. Mix a small amount of epoxy using a single drop from each tube. Put the drop into the center of the cap.

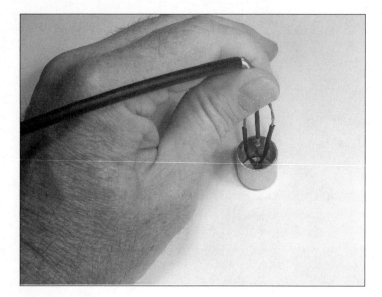

14. Place the LM34 into the epoxy with its output leads sticking straight up and the base of the LM34 pressed firmly against the inside of the cap. Hold the LM34 steady until the epoxy cures.

15. If you intend to place the sensor into liquids, epoxy the cap to the tube. Mix another small batch of epoxy and glue the plastic tubing to the cap after threading the cable through the tubing. It doesn't matter whether the cap goes inside or outside the tubing. If the tubing fits over the outside of the cap, leave about half of the cap exposed for better thermal exposure.

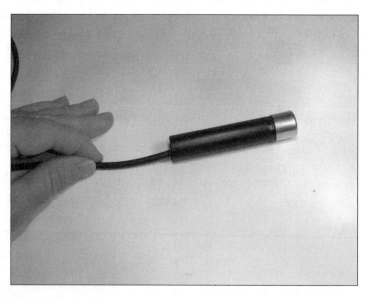

16. After the epoxy sets, slide the cap or grommet down the cable and seat it on the open end of the plastic tube. If the cable fits snugly in the grommet, you're done. If not, wrap the cable with a few turns of electrical tape, and reseat it — or you can epoxy the cable into the grommet.

17. Use the sensor as in Step 12. Take care not to get liquids into the plastic tubing — and don't heat the sensor past the point where the plastic tubing will begin to melt.

Part IV
Measuring and Testing

In this part . . .

Meet your new electronic senses! This part is a tutorial on how to use basic electronic testing instruments to make everyday measurements. You'll learn how to use a multimeter (a combination volt-ohm-current meter) for testing and checking out your circuits. Along with measuring your circuits, you have to supply them with power, too, so you'll learn to use the workbench power supply. And where do all these neat signals come from? Meet the function generator!

The last portion of this part builds on those simple measurements. Voltage, current, and resistance are all linked together by Ohm's Law, so we demystify this simple-but-powerful relationship. You also get a good working sense of the frequency and period of AC signals. With that information safely stored between your ears, you can then use decibels and measure frequency response — all this from a few simple measurements!

Chapter 11

Meet the Test Equipment

· ·

Topics in this chapter

▶ Introduction to electronic measurements

▶ Understanding about basic testing instruments

▶ Learning about advanced testing instruments

▶ Learning about the oscilloscope

· ·

No carpenter or mechanic can get by without measuring devices in the tool-box. Both have to measure length, carpenters measure angles, mechanics measure pressure, and so forth. They can't do their jobs without them and it's the same for electronic-ers like you. This chapter shows you electronic test equipment, both basic and advanced. We also introduce and explain the most marvelous piece of equipment — the oscilloscope.

What to Measure

There are four basic types of electronic measurements: voltage, current, resistance, and time or frequency. No matter what type of electronics you're working with, all of these types will be important sooner or later and sometimes at the same time! If you have some experience with electronics, you can probably skip the following discussion of each measurement.

✔ **Voltage:** This is what causes electrons to move and create current. Voltage (also called *potential*) is analogous to pressure in a water pipe. More pressure (voltage) causes more electron motion (current). Voltage, like pressure, is always measured *between* two points in a circuit. One point may be *ground* potential, using a reference connection to the Earth as one of the points. If voltage is only specified as a value, without identifying the points between which it is measured, you may assume that the voltage is measured with respect to ground or a common reference point, such as a power source's negative or neutral terminal. The units of voltage are volts, abbreviated as V. In an equation to determine voltage, the symbol E is used because voltage is a measurement of *electromotive force* (EMF).

✔ **Current:** This is the flow of electrons through a circuit, similar to the flow of water through a pipe. Unlike voltage, current is measured at a single point as the flow of electrons through some wire or contact or component. That point must be identified, such as "the current flowing into or through circuit B."

The units of current are amperes, abbreviated as amps or A. In an equation to determine current, the symbol I is used. (There are various suggestions as to why I is used, perhaps because current was termed "intensity" in early electrical experimentation.)

✔ **Resistance:** This is the resistance to current exhibited by all materials, caused by electrons moving from atom to atom or colliding with atoms inside the material. Resistance is analogous to friction in a pipe with water flowing through it. For a given pressure, higher friction reduces water flow, and for a given voltage, higher resistance reduces current. Like voltage, resistance is measured between two points in a circuit. The units of resistance are ohms, abbreviated by the Greek symbol for omega, Ω. In an equation to determine resistance, the symbol R is used, such as in Ohm's Law: $R = E / I$.

✔ **Time:** In electronics time is measured as one-time durations or delays or as intervals between repetitive events (also known as *periods*). For example, after being tripped an alarm circuit may delay for 1 second before beeping at ½-second intervals for a duration of 1 minute. The most common units of time in electronics are seconds, abbreviated as *sec* or *s*. In an equation to determine time, the symbol T or t is used.

✔ **Frequency:** The inverse of period, frequency is the number of events that occur per second. In AC circuits, the events are one complete *cycle* of the current's back-and-forth flow and frequency is measured in cycles per second or *Hertz*, abbreviated Hz. If the events are not AC cycles, frequency is given as the event name, such as marbles per second or explosions per second. The "per second" part is usually written as "/s" so frequency becomes cycles/s, marbles/s, or explosions/s. In an equation to determine frequency, the symbol f is used.

Your Basic Test Equipment

In this section, I'll introduce you to the most common and indispensable types of test equipment. Many types of electronic circuits can be tested with nothing more than these simple devices. Your goal, as an electronic-er, is to become very familiar with them, including knowing their limitations. You may get fancier or more advanced equipment later, but you have to understand these basic instruments first.

The voltmeter (okay, multimeter . . .)

Meters are everywhere. Your car has meters on its dashboard and you probably have a thermometer around your home. Anything that translates some physical measurement into a visual display on some kind of numeric scale is a meter. Meters that show voltage are common, such as the one in many vehicles that show battery voltage. In electronics, though, "multimeter" means something more.

If you go to Radio Shack and ask, "Would you show me your multimeters, please?" you will be shown several gadgets that have either a numeric display like that on your microwave oven — or a swinging needle over a collection of calibrated scales.

Not only do these devices measure voltage, but current, resistance, sometimes frequency and even more.

They're called multimeters for convenience. They are also called *voltmeters*, *volt-ohm-meters* (VOM), *digital voltmeters* (DVM), and *digital multimeters* (DMM).

Figure 11-1 shows a basic multimeter of the inexpensive type that many electronicers start with. It may be simple, but has many of the same features of its more expensive and capable siblings. Because it displays its measurement as a numeric value, it is called a *digital* meter. *Analog* meters use a swinging needle over a printed scale to display the measured value.

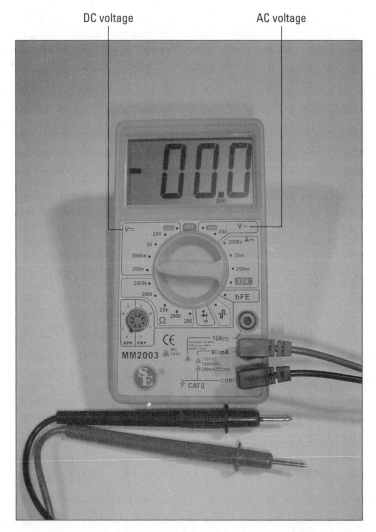

Figure 11-1: This simple multimeter is capable of measuring AC and DC voltage, DC current, and resistance over a wide range of values. Different ranges and quantities are selected by inserting the meter's probes into different jacks and setting the range switch to the appropriate scale.

In order to measure a wide variety of values, multimeters have a *range switch*. Electronic circuits involve voltages from thousandths of a volt (millivolts or mV) to hundreds or thousands of volts. Currents can range from the millionths of an amp (microamps or µV) to many amps. Resistances from tenths of an ohm to millions of ohms (megohms or MΩ) are common. Trying to measure all of these on a single meter scale would be very difficult, especially for small values. To make the meter easy to read and use in all of the different ranges of values, the sensitivity of the meter is varied. For each range and type of measurement, a different scale is used. The maximum value that can be read in each range is called the *full-scale value*.

Unless your meter has permanently attached *test probes*, there will be a set of jacks into which the probe wires are plugged. Which jacks you use determine what measurements the meter can make. The jack labeled COMMON (or COM) is used for all measurements. The black probe is usually the one inserted in the COMMON jack. For permanently attached leads, the black lead always has the same function as COMMON. There may be two voltage jacks; one for sensitive measurements of a few tenths of a volt and one for larger voltages. Some meters also have a high-voltage jack for measuring voltages of more than 500V or so.

To measure current, there must be an input and an output jack. The jack labeled "CURRENT" or "A" is the current input jack, sometimes labeled with a number to indicate maximum current, such as "10A" for a 10-ampere maximum. (Check your meter's manual to determine how your meter is labeled.) COMMON is the current output jack for all current measurements. Reversing the current flow will cause the meter to display a minus sign (–) before the value. Excessive current can damage the meter or blow an internal fuse, requiring you to open the meter and replace it.

A meter measuring current has a very low resistance between the A and COMMON jacks — almost a short circuit. This can short out a circuit when you think you're going to measure voltage. Double-check before making a measurement!

Look carefully at the AC and DC voltage labels in the figure of the meter. These international symbols for AC and DC values are used on almost all modern electronic test instruments. The AC symbol is a tilde (~) next to or under the V or A label. The DC symbol is a solid line over a dashed line, indicating that the value is of only one polarity.

Resistance is usually measured using the same jacks as voltage. Some meters use a separate jack labeled "R" or "RES." The meter applies a constant voltage to the test probes and measures current through the unknown resistance, converting the value of current to resistance using Ohm's Law (see Chapter 12). Most analog meters have a single resistance scale calibrated from zero to infinity. The range switch is labeled "×1," "×1000," "×10000," "×1M" or similarly, telling you what to multiply the meter reading by to determine the actual resistance. For example, if the meter needle is pointing to 4.7 and the range switch is set to "×1000," the resistance is $4.7 \times 1000 = 4.7\ k\Omega$.

A meter connected to measure resistance can be damaged if connected to a circuit that has power applied to it. Double-check before making a measurement!

Voltage and current are often measured using different scales for AC and DC. This is done because they are measured in different ways. Be sure to use the right scale! Many simple meters measure only DC current.

The following list shows examples of common ranges and the usual label for the switch. The numeric value of the range may be different, for example 200V instead of 100V.

- ✔ **Voltage:** millivolts (mV), volts (V), tens of volts (×10V), hundreds of volts (×100V).

- ✔ **Current:** microamps (μA), milliamps (mA), amps (A), tens of amps (×10 A or 10 A). Current range may be automatically selected by using a different jack.

- ✔ **Resistance:** ohms (Ω), tens of ohms (×10 Ω), hundreds of ohms (×100 Ω), kilohms (×1000 Ω or kΩ), tens of kilohms (×10000 Ω), megohms (MΩ).

Once the range has been selected, the meter is read by using the corresponding scale on the meter face. For example, when measuring your vehicle's battery voltage you would select the 30V range and read the voltage scale calibrated to show from 0 to 30V.

Connecting a meter set to its most sensitive range to a circuit with too much current or voltage can damage the meter by *pegging* the needle past the full-scale point. The desired meter range should be selected with the switch BEFORE the probes are attached to the circuit. If the range isn't known, the least sensitive range is selected and then increased once the probes are attached. Resistance is measured differently so an out-of-range resistance causes too little needle movement instead of too much.

Many digital meters are *auto-ranging* meters, a feature that simplifies the process of making measurements. Before displaying a value, the meter itself determines what range of values is appropriate and sets the meter display to the most appropriate scale.

Digital multimeters like those shown in Figure 11-1 perform the very same functions that analog meters do. Digital meters use a microprocessor to convert the measurement to a digital numeric value and show it on an *LCD* (liquid crystal display). Digital meters can show a measurement value with more precision than an analog meter. The microprocessor automatically calculates the correct range, eliminating the need for a range switch.

The microprocessor can also be put to work making other measurements. Digital meters are available that measure frequency and period, duration, and capacitance and inductance. Some meters can even record the measurements and download them to a computer, using a serial (RS-232) or USB port. Thus you have a simple data logger, as well!

Digital meters can also set the measurement scale automatically; this feature is called *auto-range*. All you have to do is tell the meter to measure voltage and it does the rest. Some meters can measure a value continuously and show the maximum or minimum values, a feature called *peak hold*. AC voltage and current measurements can also depend on the shape of the signal's waveform. A meter with *true-RMS* measurement capability is insensitive to the signal's shape.

When is a volt not a volt?

DC voltage is a pretty straightforward concept: The voltage has a magnitude and a polarity. AC voltage is not so straightforward; the continuous reversals of alternating current mean that the magnitude at any particular time (called the *instantaneous voltage*) is always changing. The waveform has positive peaks and negative peaks and *zero-crossings* — where the instantaneous voltage is zero — twice per cycle! So when we talk about AC voltage, what do we mean? As it turns out there are four ways of talking about AC voltage, all shown in Figure 11-2:

- ✔ **Instantaneous voltage:** The voltage at any particular instant in time, abbreviated V_{INST} or just lowercase v.

- ✔ **Peak voltage:** The maximum value of V_{INST} in any single cycle with positive and negative peaks assumed to be the same. It's abbreviated V_{PK}.

- ✔ **Peak-to-peak voltage:** The voltage between the positive peak and negative peak in any single cycle, abbreviated V_{PK-PK} or V_{P-P}.

- ✔ **Root-mean-square or RMS voltage:** The value of a DC voltage that would deliver the same amount of power to a load as the AC voltage, abbreviated V_{RMS}.

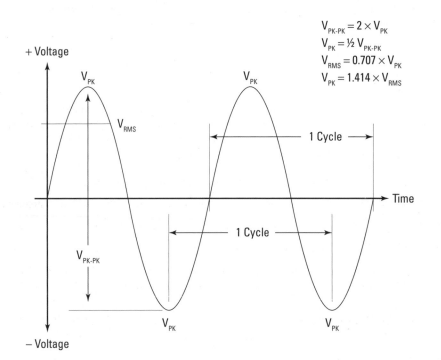

$$V_{PK-PK} = 2 \times V_{PK}$$
$$V_{PK} = \tfrac{1}{2} V_{PK-PK}$$
$$V_{RMS} = 0.707 \times V_{PK}$$
$$V_{PK} = 1.414 \times V_{RMS}$$

Figure 11-2: There are several different ways to describe the voltage of an AC waveform. Conversion factors allow you to convert one type of measurement to another for a given waveform, such as the sine wave shown here.

We were doing really well up until RMS voltage! Where did *that* come from? In the early days of electrical power, some utilities supplied DC voltage (Edison's utility was in this group) and others supplied AC voltage (Tesla's backers, Westinghouse, were in this group). Suppliers of ovens, furnaces, or motors needed a way to tell their customers how much power their products would consume or produce — on both DC *and* AC systems. Engineers determined that the RMS voltage of an AC waveform was equivalent to a DC voltage of the same value when it came to figuring out how hot a heating element would get or how hard a motor would work. In other words, if the RMS voltage of an AC waveform was the same as a DC voltage, both would do the same amount of work.

I'll not get into the gory details of the equation to determine the RMS value of a waveform. Suffice to say that for sine waves, the RMS voltage is $0.707 \times V_{PK}$. When the power company tells you that line voltage is 120V, that is the RMS value. The peak voltage is higher; $V_{PK} = 1.414 \times V_{RMS}$. For 120 V_{RMS} AC, $V_{PK} = 120 \times 1.414 = 169.7V$. The RMS-to-Peak conversions are different for other waveforms, but those are rarely needed.

This is important to voltage measurements because you need to know whether your meter displays voltage as RMS, peak, or peak-to-peak. Most multimeters read RMS by assuming the AC waveform is a sine wave and printing the corresponding scale on the meter's face, or by doing simple math in the meter's microprocessor. A multimeter that measures *true RMS* actually uses its internal microprocessor to perform the full calculation for any waveform. Check the manual for your meter to find out which type of AC measurement it performs.

The power supply

A power supply used for building and testing circuits has to be a bit more flexible than one intended for running a radio or charging a battery. Those supplies generate power at a single DC voltage that cannot be adjusted. Circuitbuilding requires a power supply with these features:

- **Adjustable output voltage:** A range of at least 3 to 15V.

- **Bipolar voltages:** Both negative and positive voltages with a common ground.

- **Meters for current and voltage:** Display the supply outputs.

- **Current:** 0.5 amps or more.

- **Voltage tracking:** The capability to match both positive and negative output voltages (for example, - (12V, (15V, (6V, and so on) is nice to have but not required.

- **Current limiting:** The capability to restrict the amount of current the supply outputs; this feature is also nice to have.

The supply in Figure 11-3 is a typical lab supply. These are available both as new and used equipment in a wide variety of sizes and power capacities. This is a *dual* supply with two independent or *floating* power supplies in the same enclosure. A meter reads the voltage and current being delivered by each supply. The terminal labeled with the ground symbol is connected to the metal enclosure or *chassis* of the supply. It can be connected to either of the power supply outputs, if need be.

Figure 11-3: A dual lab supply has two internal power supplies, completely isolated from each other and usable independently or combined.

The outputs of floating supplies (ONLY floating supplies can be used this way) can be combined by connecting the positive terminal of one supply to the negative terminal of the other, as shown in Figure 11-4. Connected together, the terminals can act as the common ground terminal of a bipolar supply, with one supply providing positive voltage and the other negative. By not connecting the common terminals to anything and taking the output from the remaining terminals, a higher voltage output than is possible from either supply because the voltages add together. Both supplies are still subject to the current limits of either supply. At high currents of several amps or more, it's best to use a single power supply.

You can use fixed-voltage supplies and even battery packs to power your circuits with the loss of flexibility and metering. With a little ingenuity, your circuits will never know! For example, a battery pack with several cells connected end-to-end can provide a different voltage from each cell; an inexpensive multimeter can act as the supply meter.

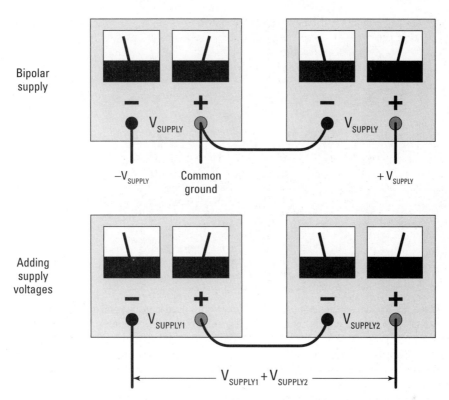

Bipolar supply

$-V_{SUPPLY}$ Common ground $+V_{SUPPLY}$

Adding supply voltages

$V_{SUPPLY1}$ $V_{SUPPLY2}$

$V_{SUPPLY1} + V_{SUPPLY2}$

Figure 11-4: Independent supplies can be combined to output bipolar voltages or add the output voltages together.

Function generators

What on Earth is a function generator? This oddly-named device generates functions, of course! (That helps, right?) Seriously, a function generator produces AC signals of various useful types — usually sine waves, square waves, and sawtooth or ramp waves.

Function refers to the type of wave (or *waveform*) and is a term from the early days of *analog computing* when equations were solved by wiring up a circuit that simulated the problem. Figure 11-5 shows the difference between the different functions. The frequency and amplitude of the output signals is adjustable; from below 1 Hz to many tens of kHz — and from millivolt levels to several volts. Function generators are used to produce test signals for circuits.

Function generators are widely available as kits as well as at any store selling used and new equipment. If you're a beginning builder, a function generator in kit form — such as the Electronic Kits FG-500 (`www.electronickits.com/kit/complete/meas/fg500k.htm`) or Vectronics VEC-4001K (`vectronics.com/products.php`) — would be a good way to learn more about building (see Part II for information about building kits) and make a useful instrument you'll be able to use for years and years.

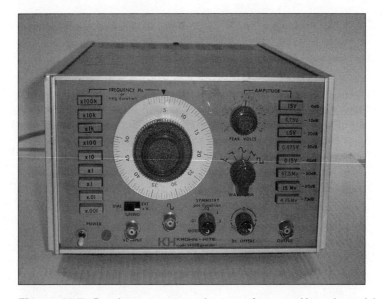

Figure 11-5: Function generators produce waveforms used in testing and developing circuits. The frequency and amplitude of the output are continuously adjustable over a wide range.

When shopping for a function generator, here are some of the features you should look for:

✔ **Types of waveforms:** Sine and square waves are a must; triangle and sawtooth are nice to have.

✔ **Waveform amplitude:** Adjustable from 10 mV$_{PK-PK}$ (or smaller) to 5 V$_{PK-PK}$ (or larger) is a good range. Somewhat less range is okay at either end of the scale.

✔ **Balanced or floating output:** As with the floating power supply, the generator's output is not connected to ground; this is handy occasionally but not necessary.

✔ **Waveform adjustments:** These are adjustments to the wave's shape; all are nice-to-haves, but not necessary:

• **DC offset:** Adds a DC voltage to the output.

• **Symmetry:** Make a square, triangle, or sawtooth waveform asymmetric.

• **Trigger:** Generates output signal only after receiving a signal from an external circuit or device.

• **Sync or logic-level output:** A square-wave signal that is the same frequency as the output waveform, used for synchronizing measurements with other equipment or as an input signal to digital logic circuits.

• **Sweep:** The ability to automatically change the output frequency through a settable range.

✔ **Digital frequency display:** Compared to a mechanical dial, it's nice to have.

Your personal test equipment — sight, sound, smell, and touch

Don't forget the test equipment that you have with you at all times: Your eyes, ears, nose and fingers are very sensitive! Even though they may not respond directly to electricity (unless you get a shock), they are quite capable of providing valuable feedback about how a circuit or device is operating.

Your eyes provide constant clues from the test instruments to the layout of the circuit. Learn to listen to that little voice of experience saying, "Something looks funny!" A test instrument giving an erratic reading, a meter giving a reading you didn't expect, a wire or probe that looks out of place — all of these are clues to troubleshooting and solving the problem.

If you're working on an audio circuit, such as an amplifier, your ears are your final Quality Control department. When you're building a project with a motor, something that's running too fast or too slow, overloaded, or just poorly controlled will make a distinctive sound. It's just like taking your car to the mechanic when it makes an unusual noise — your ears are great for detecting problems.

Who says electronic circuits have no smell? It's not true! A happy circuit board or assembled piece of equipment has a characteristic smell. When you start smelling that "hot" smell, look out! Time to switch everything off and troubleshoot. As you gain experience, you'll learn to diagnose malfunctioning circuits by aroma alone — roasted resistor, toasted transformer, charred capacitor, and melted motor all have very distinctive smells.

Be careful with touch — *never* put your fingers wherever more than 30 volts may lurk! Never! But fingers are very good at finding hot parts, even outside the enclosure. Is that exhaust air too warm? Is the case cool where it's usually hot? Is that connection loose? Did that connector take too much force to insert? Develop your own builder's touch by becoming acquainted with what properly working and assembled equipment feels like.

Use your own senses to assist you in building, testing, and troubleshooting electronics. Combine what your body tells you with what the electrical equipment tells you. Any experienced technician or engineer will have plenty of stories about how the critical bit of information came not from a multi-thousand-dollar box, but from a fingertip, scent, flicker, or crackle.

Advanced Testing Equipment

While you can get a whole lot done with the simple equipment mentioned in the preceding section, this section lists a few more instruments that can make life at the workbench a lot simpler. I advise that you get some experience with the simple equipment before jumping into the advanced list. You'll find that your specialty or interests in circuits leads you more naturally to certain kinds of test equipment.

Counters

The companion to a function generator is the *counter*. These instruments are actually not frequently used for counting. Instead, they measure a signal's frequency — counting the number of events or cycles in a certain period of time, and then calculating a frequency, just as a nurse would take your pulse in heartbeats-per-minute by counting your pulse for 15 seconds and multiplying by four. A typical portable counter is shown in Figure 11-6.

Figure 11-6: A basic counter used for field and bench measurements. Designed for use with radio signals, this counter shows frequencies as high as 500 MHz.

You'll need a counter if you need to know frequency of a signal or its period fairly precisely. Here are some examples of circuitry for which a counter would be useful:

- ✔ Tachometers
- ✔ Oscillators
- ✔ Speedometers
- ✔ Switching-type power supplies

Each of these circuits has as its input or output a sine wave, square wave, or sequence of pulses whose frequency carries information or is important in some way. You can use an oscilloscope to measure frequency as we discuss later, but for most frequency measurements a counter is much more convenient.

A basic counter should have these features:

- ✔ Counts frequency over the frequency range you're using
- ✔ Displays frequency, period (the inverse of frequency), or total (a count of cycles or events)
- ✔ Accepts signals of the type you're measuring, such as sine wave or pulses
- ✔ Accepts signals of the amplitude you're measuring, especially weak or low-voltage signals

Component testers

A multimeter can check resistors just fine, but what about capacitors, inductors, diodes, and transistors? Even if you know the component type or its *nominal* value (the labeled value), it's often necessary to test the value or to see if the component is functioning. Dedicated component testers are a worthwhile addition to your growing stable of equipment if you do a lot of building or troubleshooting.

- ✔ **Capacitor tester:** Measures capacitor value and leakage (the amount of DC current allowed between the terminals).
- ✔ **Inductor tester:** Measures the inductor value.
- ✔ **Transistor tester:** Determines whether the transistor is NPN, PNP, JFET, MOSFET and measures its current or voltage gain.

Many midrange digital multimeters have component test features built in. They can read the value of capacitors and inductors, plus check whether transistors are good and measure their current gain, or *beta*.

Logic probe

If you build a lot of digital circuits, a logic probe such as the Webtronics LP-1 (`webtronics.stores.yahoo.net/lp-1.html`) is very handy. It's a special type of multimeter, made to read the 1s and 0s of a digital circuit and display them as the color of an LED. A logic probe has clip attached to the circuit's ground; you touch the probe's tip to the circuit being tested. The LED indicates the logic level, including open circuits and active circuits switching between 1 and 0, such as an address or data bus. The fancier models can also read voltage and frequency.

Radio-frequency test equipment

This section could be a whole book by itself because there is a lot of specialized equipment for signals beyond the audio range. Special measurement techniques are required for precise measurement, too. Nevertheless, if you're just interested in checking your CB or handheld radio, a scanner, or a shortwave receiver, it's not necessary to have an engineering lab full of equipment. The following items are handy and won't break the bank:

- ✔ **Standing Wave Ratio (SWR) meter:** Measures the output power of the transmitter and whether the power is being accepted by the antenna. For an example of how to use an SWR meter with a CB radio, see `www.firestik.com/Tech_Docs/Setting_SWR.htm`.
- ✔ **Signal generator:** Works like a function generator, but outputs only sine waves and at much higher frequencies.

✔ **Dummy load:** Used when testing transmitters, dissipating the output power to avoid putting an interfering signal on the air.

✔ **Antenna analyzer:** Tests antennas and feed lines over wide frequency ranges.

✔ **Cable tester:** Checks cables to be used with radios and antennas for opens and shorts, as well as the amount of signal loss.

The Oscilloscope

The oscilloscope serves as the electronic eyes of an electronic-er, presenting visually what a signal looks like electronically. An oscilloscope (or just *scope* to those in the know) speaks volumes more about a signal than a multimeter or counter. It can also display signals at frequencies far beyond what a multimeter can handle. Learning to use a scope will take you to the big leagues of circuitbuilding!

The basics

A scope's basic function is to display the amplitude of a signal versus time. Figure 11-7 shows a scope displaying a sine wave with a frequency of 100 kHz and an amplitude of $2 V_{PK-PK}$. The signal's voltage is measured against the vertical scale and the duration of each cycle is measured on the horizontal scale. The rectangular grid on the scope's display is the *graticle*. In this case the vertical scale is 1V per rectangular division (written as 1V/div) and the horizontal scale is 5 μsec per division (written as 5 μs/div).

Graticle: The graticle of an oscilloscope is a rectangular grid of lines that makes the surface of the display look like graph paper. The graticle is divided into vertical and horizontal divisions. Depending on the sensitivity and sweep speed settings of the oscilloscope, each division can be made to represent different values of voltage or time. In that way, waveform amplitude and frequency measurements can be made.

Sweep speed

The horizontal scale is determined by the scope's *sweep speed*. The way this type of scope displays a signal is to move a beam of electrons across the face of the display tube behind the graticle, causing a phosphor coating to glow as seen in Figure 11-8. The faster the beam moves, the less time it takes to cross each division. To display higher frequency signals, the beam must move faster and the time per division will be lower. For example, to display a 1MHz signal instead of a 100kHz signal (10 times higher in frequency), the sweep speed would have to be increased from 5 μs/div to 0.5 μs/div (or 500 nsec/div). That signal would also take 2 horizontal divisions to complete one cycle as the beam moves across the tube. Adjustments to the sweep speed allow you to zoom in or zoom out to view different characteristics of the signal's behavior in time.

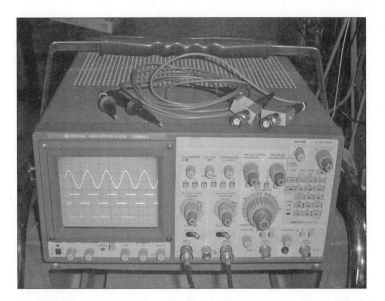

Figure 11-7: An oscilloscope displays a signal's amplitude versus time. This allows a circuitbuilder to see the exact characteristics of the signal in order to test or troubleshoot a circuit. Test probes (on top of the instrument) connect the 'scope to the circuit being tested.

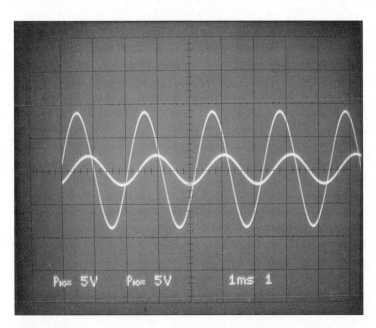

Figure 11-8: A close-up view of the oscilloscope shows the rectangular scale (the graticle) used to measure the signal on the oscilloscope's display tube. The horizontal scale is 1 msec/division and the vertical scale for both channels is 5V/division.

Vertical amplifier

The vertical scale is determined by how much the signal is amplified by the scope's *vertical amplifier*. If the gain of the vertical amplifier is increased, the signal appears larger in the vertical direction. For example, if the vertical gain in Figure 11-8 was increased by a factor of 2, the signal would be 8 divisions high instead of 4 and the vertical scale would be 0.5V/div instead of 1V/div. Many scopes have more than one input, each with its own vertical amplifier. Each input and vertical amplifier is called a *channel*. These scopes can display more than one signal at a time and at different vertical scales. Both signals share the same horizontal scale; you can compare the two different signals in detail. For example, the input and output of a filter circuit are shown in Figure 11-8 showing the different voltage levels and the delay between input and output. This is where the true power of the oscilloscope lies.

Trigger switch

The circuits in the scope that cause the electron beam to start sweeping across the display are called the *trigger* circuits. These circuits can be configured to let the beam sweep continuously, only when the signal is larger than some threshold, or just once and stop. Trigger circuits can also synchronize the electron beam to a video signal (the "TV-V" and "TV-H" settings) or to the AC power line ("Line" setting), if that helps make the displayed signal easier to see. For example, if you're trying to troubleshoot an annoying case of AC hum in a microphone amplifier, using the line trigger setting would help you see hum signals that are caused by pickup from the AC line.

Loading

Special probes are used to connect the oscilloscope inputs to actual circuits. The vertical amplifiers are very sensitive to avoid *loading* the circuit being tested. Loading occurs when a test probe or another circuit siphons off enough current to affect the circuit's operation. The most common type of probe is a *×10* (pronounced "times 10") probe. This probe reduces a signal's amplitude by a factor of 10 when connected to a scope's input. This is a result of reducing the scope's loading of the circuit. The scope's calibration and scales account for this reduction, so if you connect the circuit directly to the scope without a probe, it will appear to be 10 times bigger than it really is.

Be sure to use probes with a scope so your measurements will be accurate.

Analog and digital oscilloscopes

There are two kinds of oscilloscopes; analog and digital. Figures 11-7 and 11-8 show an analog scope. Analog scopes move an actual electron beam back and forth; they are the original and oldest type of scope. A digital scope converts the input signal to digital data and then displays it on a small computer screen — in the same form as an analog scope. Digital scopes are designed to look and function like analog scopes because circuit designers are so used to that look and feel.

The advantage of a digital scope is that the data, being digital, can be downloaded to a PC, including a screenshot of the display. A digital scope can also make automatic measurements of voltage, frequency, period, and so forth, displaying them directly on the screen. Those measurements must be made manually on an analog scope. Digital scopes also do away with the heavy and bulky display tubes of analog scopes. Nevertheless, the greater complexity of the circuitry in a digital scope drives its cost up to somewhat more than an analog scope with the same performance levels.

Making measurements with an oscilloscope

A complete description of how to measure signals with an oscilloscope is well beyond the scope of this book. There are, however, several excellent online tutorials such as the following:

Colin McCord's "Using An Oscilloscope" at `www.mccord.plus.com/Radio/oscilloscope.htm`

Hobby Project's Oscilloscope Tutorial at `www.hobbyprojects.com/oscilloscope_tutorial.html`

If you purchase an oscilloscope, the manufacturer probably includes tutorial material with the instrument or on its Web site. I highly recommend working through the tutorials in detail before using the scope to measure your circuits.

Voltage and current

Voltage is measured by displaying the signal and counting the number of vertical divisions between the points you wish to measure. If the signal occupies 5.3 vertical divisions and the vertical scale is 500 mV/div (the same as 0.5V/div), the voltage measurement is $5.3 \times 0.5 = 2.65$V. Note that the center of the graticle has extra markings to help make sub-division measurements.

Scopes have a Horizontal Position control that allow you to shift a signal back and forth on the display. Use this control to position the signal over the finer vertical scale in the center of the graticle.

Current is not directly measured by a scope and must be converted to a voltage first. A special *current probe* is used to measure current more directly. The current probe is placed over the wire carrying the current to be measured and the current's magnetic field is converted into a voltage by the probe. Current probes are fairly expensive; they're not often used in the home electronics shop.

Period and frequency

Time is measured by displaying the signal and counting the number of horizontal divisions between the points you wish to measure. If the signal repeats every 8.3 horizontal divisions and the horizontal scale is 2 ms/div, the period between signals is $8.3 \times 2 = 16.6$ ms. The frequency of a signal is the inverse of its period, so this signal has a frequency of 1/16.6 ms = 60 Hz. This signal would be a good candidate for the use of AC-line display triggering.

Specialty oscilloscopes: logic and spectrum analyzers

Logic analyzers are scopes designed for use exclusively with digital logic signals. Signals are not displayed in terms of their voltage, but rather in terms of their logic value: 0 or 1, high or low. A logic analyzer also has many channels, typically 8 to 16 at a minimum. Its test probes are connected to the analyzer as one large cable, terminating in a module with multiple thin clip leads for testing; these connect to the circuit you're testing. A logic analyzer is used to look for proper and improper operation of digital circuits, showing how the signals change at different times.

Instead of showing time on the horizontal scale, spectrum analyzers show frequency. A spectrum analyzer is really a special type of receiver. It sweeps a narrow filter across a range of frequencies, displaying the amplitude of the filter's output on the vertical scale and the filter's frequency on the horizontal scale. It's as if you rapidly tuned a radio across the band, making quick measurements of the strength of each signal. Spectrum analyzers are very, very useful to RF circuit designers and are used from audio frequencies up to many gigahertz.

Chapter 12

Measurements That Test Your Circuits and Projects

Chapter 11 looks at several different kinds of test instruments used in electronics; this chapter puts those instruments to work, assuming only that you have the basic equipment: multimeter, power supply, and function generator. The tasks in this chapter are short experiments designed to familiarize you with each technique. You'll then be ready to use your test equipment in the many ways that circuitbuilding requires.

The multimeter used in this chapter is assumed to be an inexpensive digital multimeter. These are widely available and are the usual first choice of experimenters and circuitbuilders like you. If you have a fancier model, the procedures will be the same. If you have an analog multimeter (the kind that uses a moving needle), you won't be able to read the numbers directly, but the measurements are similar. Spend some time with the operating manual for either type of meter to be sure you're familiar with its controls before beginning.

All these tasks except measuring period and frequency can be performed with advanced equipment, such as the oscilloscope. Regardless of whether you're using the basic or the advanced equipment, the stuff between your ears is the most important test instrument of all!

Making Measurements Safely

Maybe this could have been a Part of Tens chapter in the back of the book — but you need to see and consider these warnings now! Making measurements of voltage or current are done on live circuits. Although most operate from less than 30 volts with minimal hazards from shock, there may still be significant sources of energy that deserve respect and caution. This section points out ten basic safety precautions that you should learn to follow as a habit, regardless of the type of circuit:

✔ **Electricity moves a LOT faster than you can.** Don't ever think you can outrun it by swiping a finger across a terminal to see if it's hot. If it is, there's no difference between your finger contacting the electricity for 1/10th of a second or 10 seconds. The electricity will flow in nanoseconds! Use a meter or tester to make the check.

✔ **Don't give electricity a path through your body.** The most dangerous shocks occur when current flows through the chest, such as from arm to arm. Arm to leg shocks can also be dangerous. Start by not touching live conductors in the first place! Next, keep one hand in your pocket while making a measurement, if possible. Never work barefoot or in wet places or on a wet workbench.

✔ **If the equipment you're working on is powered from AC line voltage, consider using an *isolation* transformer.** Isolation transformers transfer power from the wall socket to your equipment, but without a direct connection to the AC line. The equipment is still live, but there won't be any direct connection to either the AC wiring hot or neutral wires.

✔ **Use a circuit protected by a GFCI (Ground-Fault Circuit Interrupter) circuit breaker.** A GFCI breaker can remove power from a circuit very quickly if there is an accidental short-circuit or shock hazard. This protects you when the electronics is exposed and you're making measurements. You can install a GFCI outlet for your workbench.

✔ **Don't use frayed or worn probes, cables, cords, connectors, or clip leads.** All it takes for a shock is one tiny strand of wire poking through some insulation or out of a connector. Whenever you use them, inspect them, repairing or replacing whenever necessary. Should you encounter a truly old AC line cord with cotton or rubber insulation, discard it immediately and do not try to salvage it.

✔ **Batteries as small as an AA cell can deliver enough current to melt insulation and start a fire under the right conditions.** Imagine what can happen when wiring connected to a car battery develops a short circuit or a metal tool falls across the battery's terminals! Whenever working around batteries of any size, make sure an accidental short will not blow a fuse and remove power. When working on a car's electrical system, disconnect one terminal of the battery completely.

✔ **Safety interlocks and lockouts exist for good reasons and not just to satisfy over-zealous safety regulators.** Interlocks prevent you from encountering high voltages, powerful RF fields, dangerous levels of heat, and other hazards inside equipment. Never bypass an interlock unless you're expressly instructed to do so by a manufacturer's instructions. Never leave an interlock bypassed when equipment is reassembled. If working on line-powered equipment, tag and lock out the circuit breakers so someone else doesn't accidentally turn the equipment on just as you start your repairs!

✔ **High-power radios and antennas can generate enough RF energy to cause heating of your skin and body.** The immediate effect is a burning sensation. Long-term exposure to high levels of RF energy in some frequency ranges can lead to a variety of health problems. For more information about exposure to RF, check out www.arrl.org/tis/info/rfexpose.html. Be wary of working around high-gain antennas operating above 20 MHz. If you're working on a powerful radio, consider removing all metal jewelry, particularly from your hands, since it can pick up the energy and heat or arc to your skin.

✔ **Motors and wheels and gearboxes in medium and large robots can be dangerous to the operator — that's you!** Loose clothing, ties, bandanas, long hair, or even work gloves can all be caught by a spinning shaft in an instant, especially during testing and checkout procedures. (The author has personally experienced such a scary event.) Dress properly and with respect for what a powered-up electro-mechanical system can do.

✔ **If you're working on equipment that presents a credible safety hazard, such as high voltage, make sure someone else is around to check on you.** They should also know how to remove power from your workbench outlets and perform basic first aid. CPR training would not be a bad thing for you and your friend to learn. These suggestions are not solely for electronics — they apply to any kind of workshop use.

✔ **Last, but not least, don't goof around in the shop or at your bench.** Although it may be funny to clap your hands together behind someone just about to touch a probe to a circuit, if by reacting they are shocked or damage the circuit it will quickly cease to be amusing. If you snack or sip while working (and who doesn't?), make sure the crumbs and spills can't get into your equipment or circuits.

Using Ohm's Law to Measure Resistance

A multimeter measures resistance by using Ohm's Law. The meter doesn't measure resistance directly, it measures current which it then converts to a resistance reading. Here's how:

✔ Inside the meter is a battery and a circuit that creates a constant output voltage from the battery, about 1 V or so.

✔ When the meter is set to measure resistance, it applies that voltage to the test probes.

✔ The test probes connect to the component or circuit being tested.

✔ Using Ohm's Law, the current through the probes must be I = E / R. E is the constant output voltage from the meter.

✔ By measuring the current and converting it to ohms, the meter indirectly measures resistance. As long as the test voltage is constant, this method works just fine. If the internal battery weakens and can't generate enough voltage, the meter can no longer measure resistance properly.

Another technique is to use a constant current instead of voltage. The current flowing through the resistor creates a voltage according to E = I × R. The meter converts the voltage to resistance and displays the result.

You can use Ohm's Law in just the same way when testing a circuit. For example, if you want to know how much current is flowing through a resistor, but it's not convenient to disconnect the resistor to use the meter, measure voltage across the resistor instead. Then use the equation I = E / R to calculate what the current is.

There's no place like Ohm — Ohm's Law

If you want to understand the foundation of circuits, Ohm's Law is one of the cornerstones. Georg Ohm was a German scientist that discovered the exact nature of the relationship between voltage and current in the early 1800s. With the study of electricity in its infancy, this was a major breakthrough!

Ohm found that current was directly proportional to voltage and that the constant of proportionality was a quantity he called resistance. All materials seemed to exhibit resistance and the amount of resistance varied with the type of material. Ohm stated this relationship between voltage, current, and resistance as a mathematical equation:

✔ **(a)** $R = E / I$

With a pinch of algebra, this equation can take two other forms:

✔ **(b)** $E = I \times R$

✔ **(c)** $I = E / R$

These equations "say" very simple things.

✔ **(a)** says that resistance is directly proportional to voltage given a fixed current (resistance increases as voltage increases) and inversely proportional to current given a fixed voltage (resistance decreases as current increases).

✔ **(b)** says that the voltage across a resistance is directly proportional to the current through the resistance.

✔ **(c)** says that current through material is directly proportional to the voltage across the material and inversely proportional to the material's resistance. They're all describing exactly the same relationship from differing points of view.

See the figure in this sidebar to help you remember Ohm's Law (and you should — you'll use it often). If you know any two of voltage, current, or resistance, you can find the other. Cover the unknown quantity and the figure shows you whether to multiply or divide them. For example, if you know voltage and resistance, cover the wedge labeled "I" and the two remaining wedges, E above R, show that you should divide E by R.

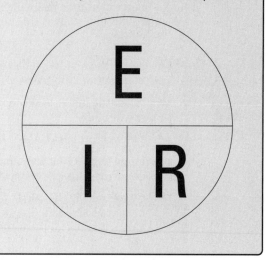

Remember these two handy sentences and you'll save yourself a lot of calculator keystrokes:

Megohms × microamps = volts

Kilohms × milliamps = volts

Both of these shortcut time-savers can be used in many common circuits.

In the tasks that follow, we'll use all three forms of Ohm's Law to ferret out the unknown value from using the two known quantities. Ready, Mr. Holmes?

Testing a Transistor

Repairing equipment and building circuits often requires testing a transistor. Is it still OK after that overload? The circuit's not working at all, but is the transistor good? Is this diode marked properly? In the "Checking a Transistor" task later in this chapter, you'll learn how to make basic checks on diodes and four common types of transistors, using only a multimeter.

Inspect your meter and determine whether it has a diode-checking function. Many digital meters do. This is usually labeled on the range switch with a diode symbol. If not, resistance measurements will be used with the meter scale set to ×1000 or a full-scale reading of 1 or 2 kΩ.

A meter with the diode-checking function tries to push a few mA of current through whatever the test probes are connected to, displaying the resulting voltage. When the test current flows through a silicon diode junction, such as is found in diodes and rectifiers and bipolar transistors, the resulting voltage drop represents the *forward voltage* of the junction. (For more information on silicon diode and transistor junctions, see the Wikipedia entry at http://en.wikipedia.org/wiki/Pn_junction.) Typical forward voltages are 0.6 – 0.75V at the current levels used by meters. The meters can also test two or three diodes in series, adding together the voltages across each junction. Check the manual of your meter to determine precisely what it can do.

Many digital meters can measure a transistor's current gain (also known as *beta* or h_{FE}). If you plan on doing a lot of circuitbuilding, this feature is often worth searching for. Keep on the lookout at flea markets or on the Web for used multimeters or kits for building transistor checkers that have actual sockets for transistor and diode leads.

Analog meters rarely have the diode-checking function, but you can still perform checks by using the meter's resistance test circuits. As you learned in the task "Using Ohm's Law", resistance can be measured by applying voltage to an unknown resistance and measuring the resulting current. When the unknown resistance is actually a silicon diode or transistor junction, the resulting current is about the same as for a 1 kΩ resistor.

Since diode and transistor junctions only conduct in one direction — from the P-type material to the N-type material — the probes must be applied to push current in that direction and the meters display results as described earlier in this section. These are called "forward tests." In the reverse direction, the multimeters will show an open circuit and these are called "reverse tests."

What if the transistor is defective or has been damaged? Junctions that are open where the tables indicate current should flow indicate a blown transistor. Transistors and diodes also sometimes fail with a short circuit and that is seen as abnormally low voltages or resistances in both directions. As you test more transistors and diodes, you'll develop a sense of what is a normal reading and what is abnormal.

Measuring in Decibels

As you gain experience with circuits and electronics, you'll discover that many measurements are made in decibels. What are these decibels and what kind of value are they? A complete discussion of decibels is beyond the scope of this book, but you can find out a lot more at the Wikipedia entry on decibels at `en.wikipedia.org/wiki/Decibel`.

The most important thing to learn (and remember!) about decibels (or *dB*) is that they are a ratio, not a unit of measure. Decibels have no units! They are a ratio of two values that both have the same units. Decibels are also logarithmic so very large ratios can be expressed in much smaller numbers as dB. You can calculate decibels as ratios of power or of voltage (or current). Here are the formulas for calculating dB where *log* is the base-10 logarithm:

$$dB = 10 \, \log\left(\frac{\text{Power 1}}{\text{Power 2}}\right)$$

$$dB = 20 \, \log\left(\frac{\text{Voltage 1}}{\text{Voltage 2}}\right) \text{ or } 20 \, \log\left(\frac{\text{Current 1}}{\text{Current 2}}\right)$$

Speaking in terms of power, a change of 10 dB is the same as a 10× change in power. Table 12-1 shows numeric and decibel ratios. If you're using voltage or current instead of power, use the second equation shown here.

Table 12-1	Numeric Ratios vs Logarithmic Ratios	
Numeric Ratio (P1/P2)	*Logarithmic Ratio = log (P1/P2)*	*Decibels = 10 log (P1/P2)*
1	0	0
2	0.3	3
4	0.6	6
5	0.7	7
8	0.9	9
10	1.0	10
20	1.3	13
40	1.6	16
50	1.7	17
80	1.9	19
100	2.0	20
1000	3.0	30

Numeric Ratio (P1/P2)	Logarithmic Ratio = log (P1/P2)	Decibels = 10 log (P1/P2)
10000	4.0	40
0.5	−0.3	−3
0.25	−0.6	−6
0.2	−0.7	−7
0.125	−0.9	−9
0.1	−1.0	−10
0.05	−1.3	−13
0.025	−1.60	−16
0.02	−1.7	−17
0.0125	−1.9	−19
0.01	−2.0	−20
0.001	−3.0	−30
0.0001	−4.0	−40

To reiterate, decibels represent quantities that are ratios of gain, loss, rejection, attenuation, and so on. Negative decibels represent a ratio less than one. If a letter follows dB, such as dBm, dbV, or dBW, that tells you what Power 2 or Voltage 2 is in the equations discussed earlier in this section.

✔ dBm means decibels with respect to 1 milliwatt (Power 2 = 1mW)

✔ dBV means decibels with respect to 1 volt (Voltage 2 = 1V)

✔ dBW means decibels with respect to 1 watt (Power 2 = 1W)

By using a specific reference power or voltage, decibels represent a specific power or voltage. For example, 10 dBm means 10 decibels greater than 1 mW or 10 mW. −20 dBV means 20 decibels less than 1V or .01V. 3 dBW means 3 dB greater than 1 watt or 2 watts. Without that letter designating a reference, decibels are ratios only without units.

Measuring Voltage

What could be simpler? Set the multimeter to VOLTS and touch the probes to the points between which you need to know the voltage! It's not quite that simple, but in this task you'll make a few measurements and pick up a few tips.

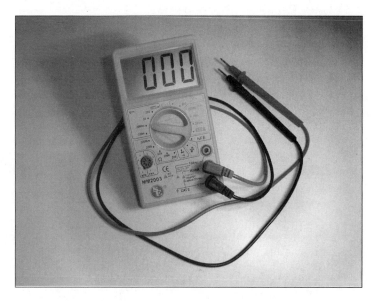

1. Inspect your meter and set it to measure DC voltage, connecting the test probes with the black probe in the COMMON or – jack. Set the range switch to the smallest scale with a full-scale voltage of 2V or higher. For example, if the meter has full-scale voltages of 2V, 20V, and 200V, use the 2V scale (sometimes labeled '2000m' for 2000 mV).

TIP

The so-called 1.5-volt battery may not be anywhere close to 1.5 volts. Rechargeable battery voltages will measure closer to 1.2 – 1.3V, depending on battery type (NiCad, LiMH, Li-Ion, and so on) or the battery's "chemistry." Alkaline cells and good old carbon-zinc batteries are the closest to 1.5V.

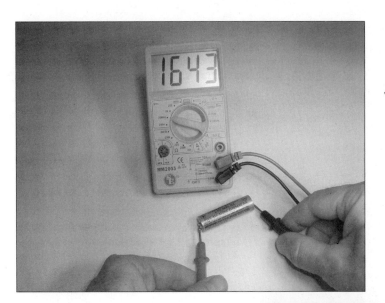

2. Place the tips of the probes on the terminals of a 1.5V battery. Determine which terminal is the positive terminal by using the multimeter. When the multimeter shows a positive value (the needles of the analog meters will move up the scale across the meter), the COMMON probe will be on the battery's negative terminal.

3. Switch the meter to measure 9V, changing the scale, if necessary. Connect the probes to the 9V battery and see which of its terminals is positive. (From here on, you're responsible for changing your meter to the appropriate scale for each step.)

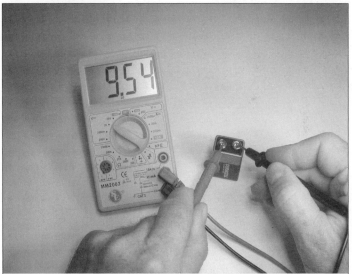

4. Measuring battery voltage when the battery is not connected to anything gives you the *open-terminal voltage*. When the battery is powering a circuit or device, its voltage will drop, much as a motor's speed drops when a load is applied. Install one of your batteries in a radio or music player and measure the voltage with the device OFF and then with the device ON to see the battery voltage change.

A 9V battery is made up of 6 small internal battery cells hooked up end to end so their voltages add together. (This is called a series connection.) Depending on the chemistry used in the cells, the sum may be anywhere from 7.2 to 9 volts.

5. Return to the 9V battery and measure its voltage. Switch the meter between reading AC and DC voltage. Does the meter reading change? Some meters will only read the AC value of any voltage, even if there is plenty of DC voltage present. These meters will read 0V for the AC measurement (or a very small value) and the battery's open-terminal voltage for the DC measurement. Other meters that can measure *RMS* (see the sidebar "When Is a Volt Not a Volt?" in Chapter 11) will measure the same value for battery voltage for both AC and DC. Check your meter's manual for a complete discussion of what the meter actually measures.

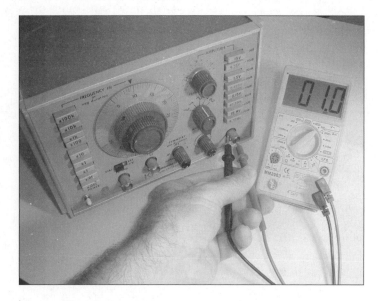

6. Turn on your function generator and set it to output a sine wave of $1V_{RMS}$ at a frequency of 1000 Hz (1 kHz). If the function generator has DC offset capability, turn it OFF. With your multimeter set to measure AC voltage, measure the output of the generator. If the generator has a coaxial connector, such as a BNC (see Chapter 8 for information on this type of connector), connect the black probe to the connector shell and the red lead to the center contact.

7. Change the meter to measure DC voltage. What happened to the reading? This waveform has no DC component, so the meter will show 0V. (If your generator has DC offset capability, turn it back on, vary the DC offset voltage and observe what happens on your multimeter.)

8. Set the meter to measure AC voltage and turn off the generator's DC offset. The generator is back to producing a 1 kHz sine wave. Measure the AC voltage at the generator's output. Without changing the output voltage, switch the generator between sine, square, and triangle or sawtooth waveforms, watching the voltage reading on the meter. Waveform shape makes a difference in how your meter measures AC voltage as described in Chapter 11.

Components: A voltage can have both AC and DC characteristics, called components. For example, a power supply may output 12V DC, but with small AC variations around 12V called ripple.

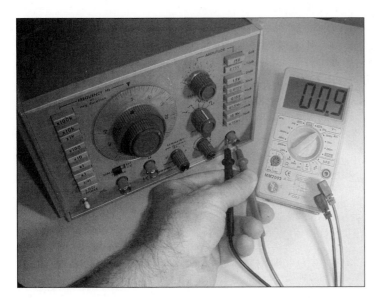

9. Return the generator's waveform to a sine wave. Check the meter's ability to measure AC voltages at higher frequencies by increasing the generator's frequency until the AC measurement begins to fall. Where the meter's reading drops by more than a few percent is the upper limit of frequencies at which the meter's reading is correct. Readings at higher frequencies will be lower still, causing the meter to read erroneously low values.

10. If you constructed any of the circuits in Part II or have other projects lying around, take this opportunity to exercise your newfound measurement skills. With the stereo pumping out some music, measure the voltage at the speaker terminals. What is the output voltage from your collection of "wall wart" power supplies that plug into the wall socket? (Be sure to measure both the AC and DC components!) Measure the voltage on your car's battery with the engine off and with the engine running.

11. Now for the main event! A drum roll, please . . . measuring AC voltage directly from a wall socket! This is completely safe if your test probes are in good condition as discussed in the safety list at the beginning of this chapter and you have the meter set to the right range for measuring 115V AC. After checking both of those things, approach the wall socket. Keeping your fingers well away from the test probe's metal tips or clips, stick the black probe into one of the plug slots (it doesn't matter which). Now stick the red probe into the other slot. Your meter will now read somewhere between 100 and 130V AC, depending on how the power company is doing that day. Remove the probes from the slots and mop the sweat from your brow. Good work!

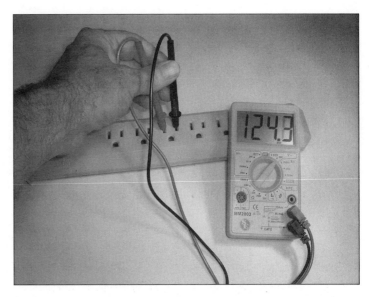

Measuring Current

Measuring current is a lot like measuring voltage except that the current must be routed *through* the meter. You must make the same considerations for AC and DC currents; moving the test probes between jacks and changing the meter's range switch to the proper position. Current flows into the meter through the jack labeled "Current." Current flows out of the meter through jack labeled "COMMON" or -.

In this task, you'll measure current flowing in a light bulb, powering a music player, and in a simulated speaker. These measurements are typical of those you'll perform as part of building circuits and you'll gain experience in setting up your meter to measure current.

Whenever changing a multimeter between measuring voltage and current, be sure to insert the test probes into the correct jacks. The resistance through the meter when measuring current is very small, so having the test probes in the current jacks when you connect them to measure voltage can short out the circuit you're testing. This may blow a fuse inside the meter or damage your circuit, so be careful!

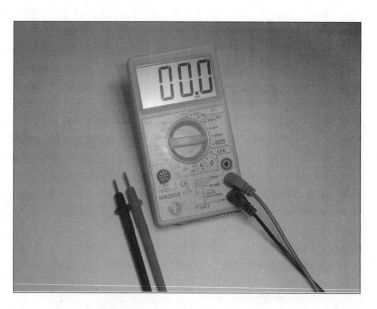

1. Inspect your meter and set it to measure current. If there is a choice, select DC current. Insert the test probes into the current output (COMMON or -) and current input jacks. If your meter has different jacks for different ranges of current, use the jacks for currents of approximately 200 mA.

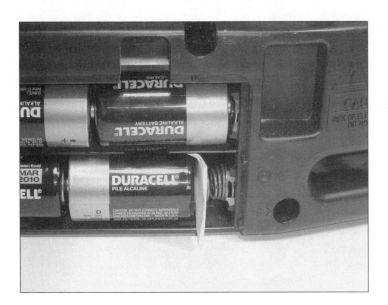

2. Turn the radio or music player OFF and open the battery compartment. Push the battery away from the battery holder contact and slip in the strip of paper as shown in the figure. This opens the circuit between the batteries and the internal electronics.

3. Connect one of the meter's probes to the battery holder contact on one side of the paper strip, using a clip lead if necessary, or just slipping the probe into the spring as shown in the figure. Slip the other probe between the paper strip and battery terminal.

4. With a speaker or headphones attached, turn the radio or music player ON, tune in a station or play a song and observe the meter. If your meter has more than one current scale, select the smallest scale whose maximum value is larger than the displayed current. For example, if the current is 25 mA and the current scales on the meter are 2000 mA (2 A) and 200 mA, use the 200 mA scale.

5. Turn the volume up and down to see what happens to the current being drawn from the batteries. If you can switch the player between a speaker and headphones, see which draws the most current from the batteries.

Measuring Resistance with Ohm's Law

Stuff You Need to Know

Toolbox:

- ✔ Analog or digital multi-meter with test probes
- ✔ Clip leads
- ✔ Power supply

Materials:

- ✔ 100 Ω, 1 kΩ, 10 kΩ, ¼-watt resistors

Time Needed:
About an hour

If you're unfamiliar with Ohm's Law, please start this task by reading the sidebar, "There's No Place Like Ohm — Ohm's Law" earlier in this chapter. Whether you design your own circuits or not, understanding Ohm's Law is an important part of electronics testing and troubleshooting.

1. Inspect your meter and set it to measure resistance. Select the full-scale range closest to, but greater than 1 kΩ. With the probes separated, your meter should indicate an open circuit — usually a "1" and no other digits. With the probe tips touching each other, the meter will display a value of a fraction of an ohm.

2. Measure all three resistors one at a time by applying the test probes to each resistor's leads. To measure the 100Ω and 10kΩ resistors, select smaller and larger full-scale ranges, respectively.

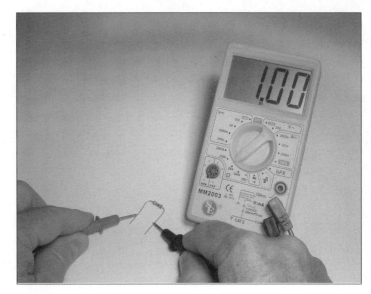

3. Connect all the resistors *in series* by using a solderless breadboard (as shown in the figure; see Chapter 3 for more information), soldering them, using clip leads, or just twisting their leads together.

In series: A method of connecting electrical components end-to-end so there is only one current path and the same current flows through all the components being connected together.

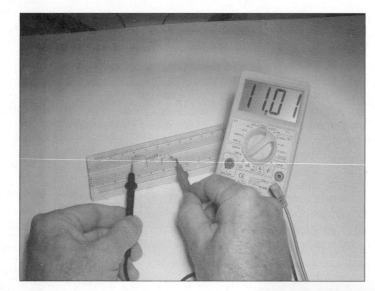

4. Measure the resistance from one end of the resistor string to the other. You add the values of resistances in series together, so the total resistance should be 100 + 1000 + 10000 Ω = 11100 Ω. It may be a little more or less because the labeled resistance values aren't exact. (See Chapter 1 for more information on component tolerances.)

5. Now connect all the resistors together *in parallel*.

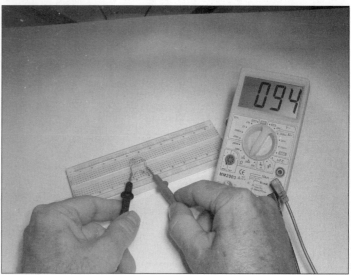

6. Measure the resistance of the parallel set of resistors. The reciprocal of the sum of the reciprocals (got that?) is how you add resistance values in parallel. Calculate the expected parallel value by taking the reciprocal of each resistor's value, for example, 1/100, 1/1000, and 1/10000. Add these together, for example, .01 + .001 + .0001 = .0111. Now take the reciprocal of that: 1/0.0111 = 90.09 Ω. Compare your measured and calculated values.

GLOSSARY

In parallel: A method of connecting electrical components with one set of leads tied together and the other set of leads tied together so when voltage is applied, the same voltage is applied to all three resistors. In parallel circuits, the current must divide between the different branches of the circuit.

7. Moving on to Ohm's Law, turn your power supply on and adjust it to 12V output or prepare a battery pack with about the same voltage. Measure the exact value of the 1 kΩ resistor with your multimeter.

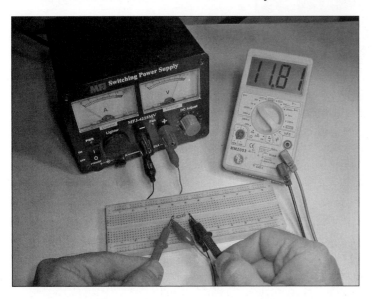

8. Using clip leads, connect the 1 kΩ resistor between the power supply (or battery pack) + and – terminals, then turn the supply on. Measure the voltage across the resistor. You know voltage (12V) and resistance (the value you measured in Step 7), so use the equation I = E / R to calculate the current flowing through the resistor. If the voltage is exactly 12V and the resistor value was exactly 1 kΩ, the current would be 12 / 1000 = 12 mA.

9. Connect all three resistors in series again as in Step 3. Apply voltage from the power supply across the ends of the string. Use Ohm's Law as in Step 8 with the total resistance you measured in Step 4 to determine the current flowing in the series string. If your supply voltage is 12V, the current should be close to 12 / 11100 = 1.1 mA.

10. Now use Ohm's Law to determine what voltage you should measure across each resistor. You know the value of each resistor and the value of the current flowing through the string. Since the same current flows through *all* the resistors (where else can it go?), the voltage across the resistor is E = I × R. If 1.1 mA is flowing through the 1 kΩ resistor, the voltage across it should be

1 kΩ × 1.1 mA = 1.1V.

(Remember: kilohms × milliamps = volts.)

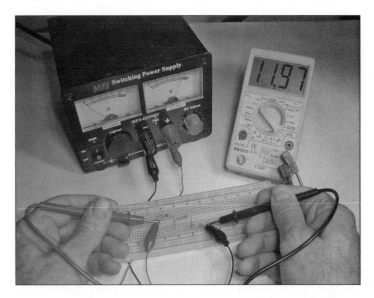

11. Repeat Step 10 for the 100 Ω and 10 kΩ resistors. What voltage did you calculate for each resistor? (The voltages should be 0.11V and 11.1V, respectively).

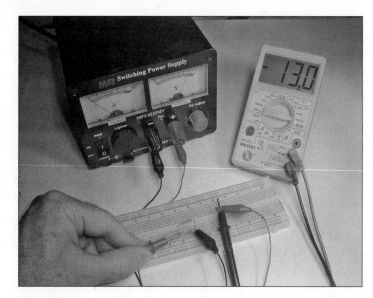

12. Connect just the 1 kΩ and 10 kΩ resistors in parallel as in Step 5 and measure their resistance as in Step 6. Set up your meter to measure current and connect the meter so current from the power supply's + terminal flows through the meter before flowing through the paralleled resistors.

13. Calculate the total current you expect to flow through the paralleled resistors from a 12V supply using the value for the resistors in parallel from Step 6. If the voltage is 12V and the parallel resistors measured 909 Ω, then I = E / R = 12 / 909 = 13.2 mA.

14. Apply 12V from the power supply and measure the actual value with your meter. You have now used Ohm's Law in all three ways!

Checking a Transistor

Stuff You Need to Know

Toolbox:
- Analog or digital multimeter with test probes
- Clip leads

Materials:
- 1N4148 or 1N4001 diode
- 2N3904 transistor
- 2N3906 transistor
- 2N7000 transistor

Time Needed:

About an hour

This set of tasks covers the three types of transistors you're most likely to encounter, but there are many, many others. Use the Web resources and your experience to determine how other types may be checked with your multimeter.

Table 12-2 lists test results for the following steps. Tests for meters with the diode-checking function are in the column labeled "DF Result." Tests for analog meters using resistance checks are in the column labeled "RM Result." A red probe is assumed to be connected to the meter's voltage/resistance measurement jack and a black probe to the COMMON jack. Table 12-2 lists each test connection in the red-to-black direction. For example, in the measurement tables, "Anode to Cathode" means to take a measurement with the red probe on the diode's anode lead and the black probe on the diode's cathode lead.

1. Download the data sheets for each of the diodes and transistors by entering the part number and the words "data sheet" into an Internet search engine. For example, *1N4148 "data sheet."* Several sources will be listed for each part.

TIP

Create a Data Sheets folder on your computer. Each time you look up a part's data sheet, download the PDF version and store it in this folder or use PDF printer software to create a PDF version. (If the file is not clearly named, change the name to *part number* data sheet.) This way you'll build up a library of data sheets you can access any time.

MOSFETs: The next generation

A *MOSFET*, or Metal-Oxide Semiconductor Field-Effect Transistor is an improved version of the bipolar transistor, using less power to operate. Instead of the bipolar transistor's base-collector-emitter, the corresponding MOSFET terminals are the gate, drain, and source. MOSFETs also have two different types of drain-source channels — N-type and P-type like bipolar transistors — and they can either turn on with increasing gate-to-source voltage (enhancement-type) or turn off (depletion-type). (For more information about MOSFETs, check out the Wikipedia entry at `http://en.wikipedia.org/wiki/Mosfet`.)

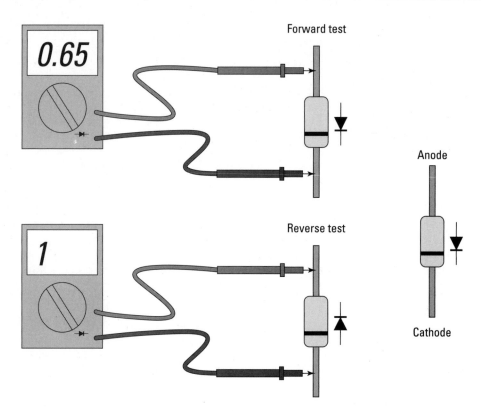

2. On a 1N4148 diode or 1N4001 rectifier, determine which lead is the cathode by looking for the painted bar at one end. Make the tests as shown in Table 12-2.

Table 12-2	Diode Junction Tests	
Test Connection	*DF Result*	*RM Result*
Anode to Cathode	0.6 to 0.75V	500 Ω to 2 kΩ
Cathode to Anode	Open	Infinite

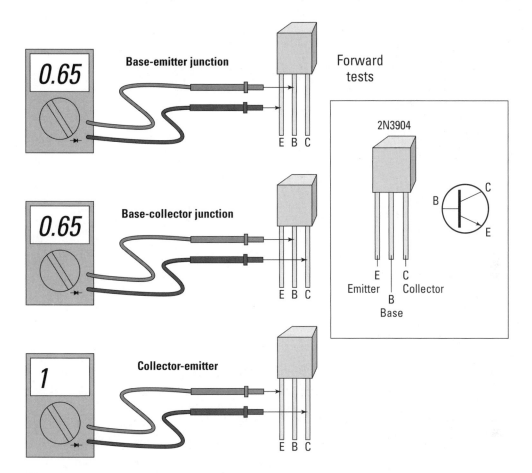

3. Determine which leads of the 2N3904 transistor are the collector, base, and emitter. The 2N3904 is an NPN device, meaning the collector and emitter are N-type material and the base is P-type material. Therefore, current will only flow from the base to the collector and from the base to the emitter. Make the tests as shown in Table 12-3. (The figure only shows the forward tests.) The most common failure of these small bipolar transistors is from overload, causing low resistance readings (below 500 Ω) across junctions or making junctions act like resistances in both directions.

Table 12-3	NPN Transistor Junction Tests	
Test Connection	*DF Result*	*RM Result*
Base to collector	0.6 to 0.75V	500 Ω to 2 kΩ
Collector to base	Open	Infinite
Base to emitter	0.6 to 0.75V	500 Ω to 2 kΩ
Emitter to base	Open	Infinite
Collector to emitter	Open	Infinite
Emitter to collector	Open	Infinite

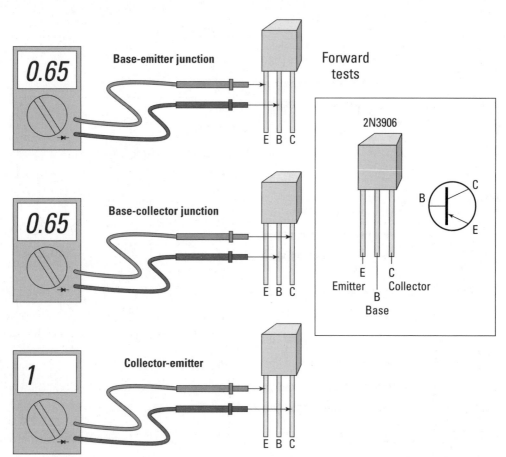

4. Determine which leads of the 2N3906 transistor are the collector, base, and emitter. The 2N3906 is a PNP device, meaning the collector and emitter are P-type material and the base is N-type material. Therefore, current will only flow from the collector to the base and from the emitter to the base. Make the tests as shown in Table 12-4. The most common failure of these small bipolar transistors is from overload, causing low resistance readings (below 500 Ω) across junctions or making junctions act like resistances in both directions.

Table 12-4	PNP Transistor Junction Tests	
Test Connection	*DF Result*	*RM Result*
Base to collector	Open	Infinite
Collector to base	0.6 to 0.75V	500 Ω to 2 kΩ
Base to emitter	Open	Infinite
Emitter to base	0.6 to 0.75V	500 Ω to 2 kΩ
Collector to emitter	Open	Infinite
Emitter to collector	Open	Infinite

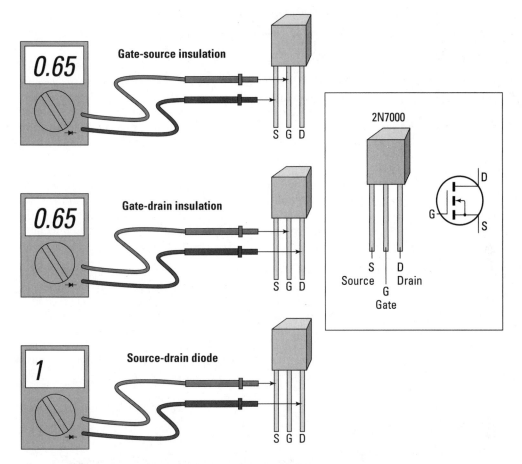

5. The 2N7000 is the most common type of MOSFET as described in the sidebar; an N-channel enhancement MOSFET. As Table 12-5 shows, MOSFETs are tested differently than bipolar transistors because the gate is completely insulated from the drain and source so no current flows in the gate terminal. Due to their construction, MOSFETs have a built-in source-to-drain *body diode* that tests like a regular silicon diode. There are two common failures of small MOSFETs. *Gate punchthrough* breaks down the gate insulation resulting in low resistances (less than 10 kΩ) between the gate and drain or source. Overload can cause the drain-to-source resistance to be low (less than 500 Ω in an enhancement mode transistor).

Table 12-5	N-Channel Enhancement MOSFET Tests	
Test Connection	*DF Result*	*RM Result*
Gate to source	Open	Infinite
Source to gate	Open	Infinite
Gate to drain	Open	Infinite
Drain to gate	Open	Infinite
Drain to source	Open	Infinite
Source to drain	0.6 to 0.75V	500 Ω to 2 kΩ

Measuring Period and Frequency

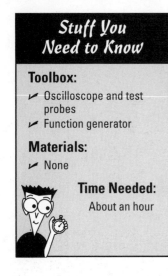

For this task, you won't be able to use a multimeter (unless it's one of the more capable digital meters with a frequency counter built-in). As a result, this task uses an oscilloscope. If you don't have an oscilloscope of your own, ask around and see if someone you know has one or knows someone that does or has access to a scope. Many hobbyists have one!

Before beginning this task, familiarize yourself with oscilloscopes in general by reading through the tutorials listed in Chapter 11. Your scope may not have exactly the same control arrangement or labeling, but all of them have the same basic features that you'll use in this task.

1. Inspect your scope and find the calibration output. It should be on the front panel and may be labeled "CAL." On most scopes it's a small, exposed terminal or jack. The CAL signal is usually a 1 kHz square wave with an amplitude of $1V_{PK-PK}$ or less. You use the CAL signal to check the scope display and adjust the test probes as described in the scope manual. (If your scope doesn't have a CAL signal output, use your function generator to produce a 1 kHz, $1V_{PK-PK}$ square wave.)

2. Set up the scope for a vertical gain of 0.5 V/div (use Channel 1 or A if it's a multiple-channel scope), and a sweep speed of 1 ms/div. Set the trigger mode to AUTO. You should see a bright line or *trace* on the display. Adjust the Vertical Position control so the trace is on the graticle at the center line. Adjust the Horizontal Position control so the trace is approximately centered left-to-right. (If you can't see the trace, most scope manuals include a procedure for setting all controls so a trace should be visible. There are a lot of settings on a scope and some of them can move the electron beam off-screen or prevent it from sweeping across the display.) Adjust the Focus control for the sharpest, narrowest trace. Adjust the Brightness or Intensity control so the trace is easy to see on the display, but doesn't blur.

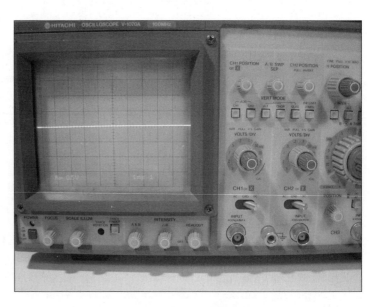

3. Connect a test probe to Channel 1 and attach the end of the probe to the CAL terminal or jack. You should now see a square wave on the display instead of the straight line trace. If the trace disappears, you may have the vertical gain improperly set. Reduce vertical gain (set the control to higher values of voltage per division) to see if the waveform appears.

4. Determine the amplitude of your scope's CAL signal by counting the vertical divisions between the square wave's maximum and minimum voltages. You can make a more accurate count by using the scope's Horizontal Position control. Move the waveform's upper and lower portions so they are next to or cross the more finely graduated center vertical axis. Multiply by the vertical gain in volts per division. This gives you the peak-to-peak voltage of the waveform.

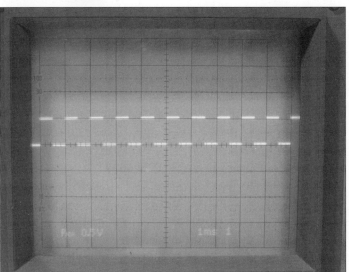

5. Repeat your amplitude measurement at higher and lower vertical gain settings, making the waveform smaller and larger. When the waveform is small, the width of the trace itself can introduce error into the measurement: It's hard to tell if you measure at the top, bottom or center of the trace. Amplitude measurements are best made with the waveform as large as it can be and still be completely within the graticle. This minimizes the effect of the trace width on the measurement.

If you want, repeat each amplitude measurement that follows with your multimeter to get a feel for how multimeter and scope voltage measurements compare.

6. Determine the frequency of your scope's CAL signal by counting the horizontal divisions between the square wave's low-to-high edges or *transitions*. Use the Horizontal Position control to move the left-hand edge to a vertical line on the graticle. The center horizontal axis is your measuring scale. Multiply the number of divisions by the horizontal sweep speed in time per division to get the period of one cycle of the waveform. Its frequency is the reciprocal of period (1/period). (If you have a frequency counter, connect it to the CAL output and compare its reading with yours.)

7. Repeat your frequency measurement at faster and slower sweep speed settings, stretching and compressing the waveform horizontally. As with amplitude, the most accurate measurements of period are made with the waveform as large as it can be and still have one complete cycle completely within the graticle.

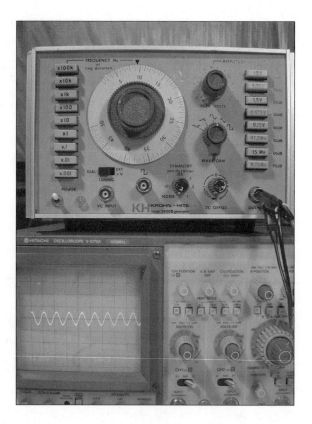

8. Reconnect the test probe to the output of your function generator. Replicate the CAL signal with the generator. For example, a square wave with the same amplitude and frequency. Confirm that you have the same type of waveform and displayed trace.

9. Switch the generator to output a sine wave without changing the amplitude or frequency. How can you measure frequency and amplitude now? Measure amplitude by using the Horizontal Position to move the waveform maximum and minimum points exactly on the center vertical axis to count divisions. Since sine waves don't have sharp vertical edges, use the points at which the waveform crosses the center horizontal axis as your measurement points.

10. Try other waveforms that your generator can produce to practice making amplitude and frequency measurements on those shapes. You'll find that each shape requires a slightly different technique to count vertical and horizontal divisions accurately. The key to making good measurements with a scope is being able to view a stable trace and positioning it correctly on the graticle. By practicing, you'll become skilled in applying this powerful "electronic eye" to your circuitbuilding activities.

Making Measurements in Decibels

Stuff You Need to Know

Toolbox:
- Analog or digital multimeter with test probes
- Function generator
- Clip leads or generator output cable with test clips

Materials:
- 100 Ω, 1 kΩ, 10 kΩ, ¼-watt resistors
- Make a copy of the decibel graph to use as graph paper

Time Needed:
Less than an hour

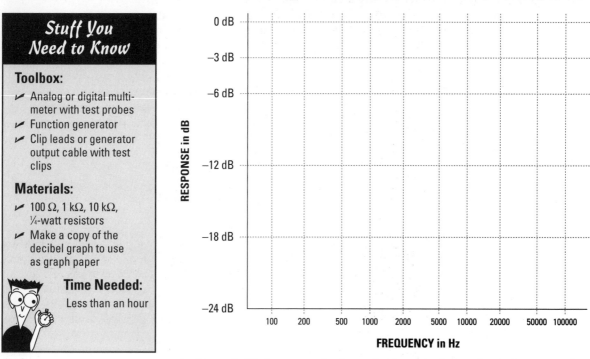

In this task, I'll show you the formulas for calculating decibels and you'll get some practice in converting voltage measurements to decibels.

This graph shows decibels (dB) on the vertical axis and frequency on the horizontal axis. By plotting a circuit's output-to-input ratio in dB at different frequencies, the graph shows frequency response. Frequency is shown logarithmically to show equal ratios of frequencies as equal space along the axis. This gives a better picture of how the circuit responds than if frequency was shown linearly.

1. Inspect your meter and set it to measure AC voltage.

2. Set the function generator to output a 1 kHz sine wave of $2V_{RMS}$ amplitude.

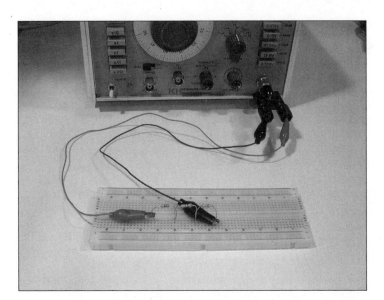

3. Connect the three resistors in series (see the task "Measuring Resistance with Ohm's Law" earlier in this chapter). Use clip leads to connect the string of resistors to the output of the function generator.

4. Measure the voltage across the 10 kΩ resistor. It will be about 90% of the total voltage from the generator. Using the generator output of $2V_{RMS}$ as Voltage 2, calculate the attenuation of the voltage across the 10 kΩ resistor using the equation dB = 20 log (measured voltage / $2V_{RMS}$). The result should be about –0.83 dB.

5. Measure the voltage across the 1 kΩ resistor and perform the same calculation. The result should be about –20.9 dB.

6. Measure the voltage across the 100 Ω resistor and perform the same calculation. The result should be about –40.9 dB.

7. What is $2V_{RMS}$ in dBV? Use $1V_{RMS}$ as Voltage 2 in the equation.

$$dB = 20 \log\left(\frac{2V_{RMS}}{1V_{RMS}}\right). \ 2V_{RMS} = 6 \ dBV.$$

8. Convert each of the voltages you measured across the resistors to dBV using the equation dB = 20 log (measured voltage / 1 V_{RMS}). The answers are 5.2 dBV, –14.8 dBV, and –34.9 dBV.

Measuring Frequency Response

One of the most closely read specifications (or "specs") of home audio equipment is its frequency response. For example, an amplifier may be rated at full output from 20 Hz to 18 kHz. What does that really mean? When discussing an amplifier, it means that below 20 Hz and above 18 kHz, the amplifier's output power will be less than half of the output *between* 20 Hz and 18 kHz.

Frequency response is *relative;* that is, it's related to some reference measurement or level of performance. Because frequency response is relative, it's measured in dB. In the amplifier example, frequency response shouldn't depend on how loud the output is or isn't. By using dB, the measurement is made to be independent of output volume.

Frequency response is just one of many types of measurements that use dB to give relative results. In this task, you'll make a set of measurements on a simple circuit that show you how to combine frequency measurement, voltage measurement, calculate dB, then graph the result in the standard way that other circuitbuilders will understand.

1. Inspect your meter and set it up to measure AC voltage. Set your function generator to output a sine wave, at 100 Hz, of about 2V$_{RMS}$. (The exact voltage isn't critical, as long as it isn't changed during the task.)

2. Build the circuit shown in the figure. You can use clip leads and twist component leads together if you want. A solderless breadboard is a little easier to use. Use the function generator's output connector shell as the circuit ground.

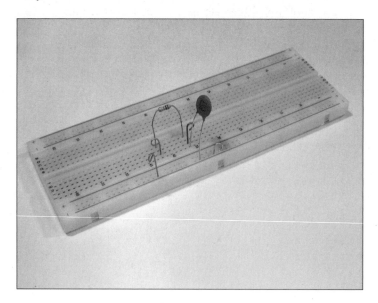

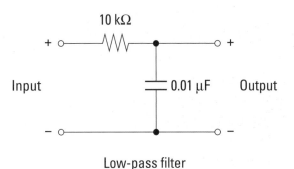

10 kΩ

Input   0.01 μF   Output

Low-pass filter

This circuit is a *low-pass filter* — its output gets smaller and smaller as frequency increases because at higher frequencies the capacitor becomes closer and closer to a short circuit. Filters made from resistors and capacitors are called *RC filters*.

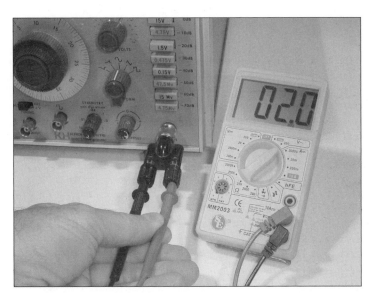

3. Turn on the function generator and measure its output voltage directly at the output connector. Record this voltage — it's the reference voltage for all dB calculations.

4. Reconnect the multimeter to the capacitor, one probe on each lead. You should read approximately the same voltage on the multimeter.

5. Set the function generator's frequency to all the frequencies in Table 12-6, taking a voltage measurement at each one.

Table 12-6	Frequency Response Measurement Table	
Frequency in Hz	*Voltage in V_{RMS}*	*Response in dB*
100	X	0.0
200	X	X
500	X	X
f_0	X	−3.0
1000	X	X
2000	X	X
5000	X	X
10000	X	X
20000	X	X
50000	X	X
100000	X	X

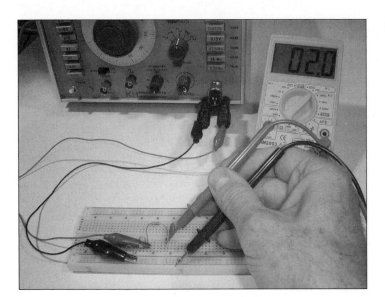

6. Readjust the frequency until the multimeter reads 0.707 times the reference voltage you measured in Step 3. This should be around 1600 Hz for the values of resistance and capacitance in the circuit. (Any frequency between 1200 and 2000 Hz is reasonable.) This frequency is called the filter's *cutoff* frequency or *corner* frequency because this is the point at which the filter's output power is half that of the input power from the signal generator.

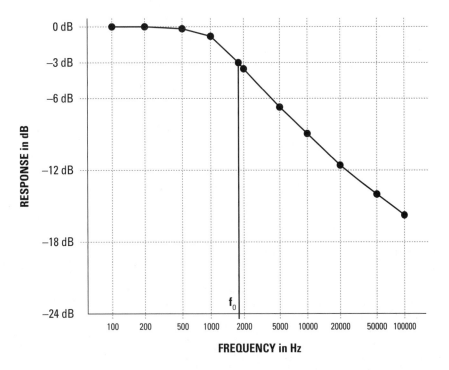

7. Calculate the dB response for each frequency and plot all the points on the graph paper. Use a ruler to draw a line from the lowest-frequency point to the next lowest. From that point to the third-lowest, and so forth, eventually creating a curve that describes the filter's frequency response. Draw a vertical line from the point located at the cutoff frequency to the frequency axis. Label that point "f_0". You've just drawn a frequency response curve!

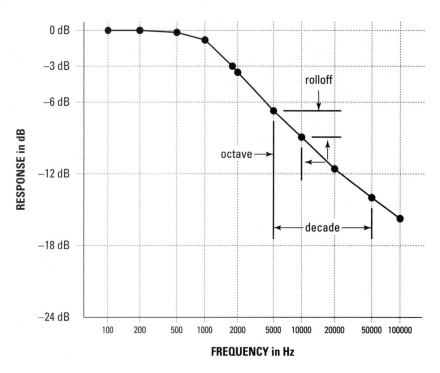

8. Above f_O, the response curve quickly becomes a straight line sloping down as frequency increases. Subtract the dB response at 10 kHz from the dB response at 5 kHz. This is one *octave* of difference in frequency. The difference in dB over a change of one octave is the filter's *rolloff* in dB/octave. Subtract the dB response at 50 kHz from the dB response at 5 kHz. This is one *decade* of frequency difference and the rolloff is in dB/decade. Both figures are often used to describe a filter's frequency response.

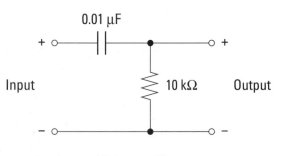

High-pass filter

9. While you have the equipment all set up, change the filter to a *high-pass* filter by exchanging the positions of the resistor and capacitor. For example, the capacitor is connected to the function generator's output, then to the resistor, which is then connected back to the output connector's shell. Repeat the measurements and graph the results.

You can also design your own simple low-pass and high-pass RC filters using the following equation:

$$f_0 = \frac{1}{(2\pi\,R\,C)} = \frac{0.159}{RC}$$

Experiment by changing the value of the resistor and the capacitor, then finding the new value of f_0 both by calculation and by measurement.

Part V
Maintaining Electronic Equipment

The 5th Wave   By Rich Tennant

"So, breadboards and power drills...
not a good mix, my friend."

In this part . . .

When you've finished building a circuit, the fun is often just beginning! As you build more and more circuits, you'll find that your experience lets you work with equipment of all sorts. In this part, we touch on techniques for maintenance and troubleshooting that you'll find very handy in the days of circuitbuilding that lie ahead for you.

Chapter 15 is devoted to that ubiquitous energy source, the battery. With so many types and styles available, how do you choose just one — and why do they perform the way they do? Read and behold the secrets revealed!

Electronics are seemingly everywhere and that extends to vehicles. Your car is full of electronics and opportunities abound for adding more. The automotive environment presents its own special challenges, so we'll review the basics of mobile installation. With electronics cropping up everywhere, all that equipment and all those gadgets may interact in ways not intended by their designers! Chapter 17 presents some techniques for dealing with noise to (and from!) your circuits.

Chapter 13

Who Let the Smoke Out?

So far, this book has been about learning the skills of building circuits. There's more to building than just the mechanical construction, of course. Unless you're very lucky (or very, *very* good), there will be times when that circuit or project or system that you just assembled doesn't work. What then? This chapter shows you how to cope.

It's a little-known secret (kept in confidence by the experts) that all electronics works because of the Magic Smoke inside every component. If you let the Magic Smoke out, it stops working, it's that simple! (Well, at least nobody's been able to prove otherwise.)

In these few pages, it won't be possible to give you step-by-step procedures for every contingency. In fact, that's not possible even in hundreds of pages! Every circuit and piece of equipment is a little bit different in the problems that beset it and how it acts in response. The goal of this chapter is to extend your circuitbuilding skills by introducing some basic methods that you can apply to *all* those different problems. After you get good at using them, you'll be pleased to learn how quickly the problems yield to your awesome tool kit!

Troubleshooting and Debugging Basics

Let's start at the beginning, shall we? *Troubleshooting* is what you do when something has failed. This implies that at some point the device has done what it's supposed to. *Debugging* is what you do when the device hasn't yet demonstrated that it's capable of doing what it's supposed to do. In both cases, the device is in a state of failure. (For the rest of this chapter, I'll use the term "troubleshooting" to include "debugging.")

What is failure?

What is *failure?* (Technically speaking, of course.) According to the American Society of Civil Engineers, "Failure is an unacceptable difference between expected and observed performance." Failure can be as simple as a power supply that won't turn on or as complex as a microprocessor circuit with occasional glitches. There are three parts to this definition of failure:

- ✔ **Expected performance:** This is what the device is supposed to do. You have to be able to state clearly what the expected performance is. Otherwise you can't say whether it's broken *or* working!

- ✔ **Observed performance:** This is what the device is actually doing. You also have to be able to say *what* it's doing; otherwise you can't really say how it failed!

- ✔ **Unacceptable difference:** No device is perfect, so it will perform slightly differently than however it's specified to perform. Only when the discrepancy is large enough does it become failure.

Before you embark on those voyages of discovery we call troubleshooting, take some time to think about those three ideas. If they're complicated enough (or the device is complex enough), write down your thoughts. You need to begin with a good idea of why you're troubleshooting and what you intend to accomplish — even for simple repairs and tests.

After your acumen in electronics becomes known, you'll no doubt be visited on occasion by family and friends bearing gadgets and items in need of attention. When you ask what's wrong, the reply will be, "It's broken!" At this point, dialogue must take place as you extract the details of what "broken" means. A crack or fracture? Does it not turn on? Did the Magic Smoke get out? Use your best manner of interrogation to extract the details — because you'll need them!

Running in circles

Troubleshooting can feel a lot like running in circles — because it is! You start by noticing some kind of failure and then you start looking for the cause. Around and around you go! Eventually, the trouble is found (or you give up) and the thing gets fixed (or it doesn't). Isn't there a better way? Sort of. Figure 13-1 shows the game board on which troubleshooting is played.

- ✔ **Assess:** Take the time to carefully observe the symptoms. This could mean turning it on and taking notes or it could mean taking detailed measurements.

- ✔ **Compare:** Compare your observations to what the device is supposed to be doing. This could be described in a specification or maybe a user's manual. Be specific and try not to assume or guess.

- ✔ **Hypothesize:** This really means that you think of reasons why the device could behave the way it does. Pick one reason and come up with measurements you could make to prove or disprove the reason. For example, "The output amplifier IC could be distorting because the power supply voltage is too low."

✔ **Measure:** Go take those measurements so you can tell if your hypothesis is correct.

✔ **Make a change:** If measurements show your hypothesis is correct, make a change to clear up the problem, such as an adjustment or replacing a component. If the hypothesis was not correct, compare again and come up with another hypothesis.

✔ **Compare again:** After the change has been made, assess the symptoms again to see if they have gone away. If so, it's fixed — congratulations! If not, go around again!

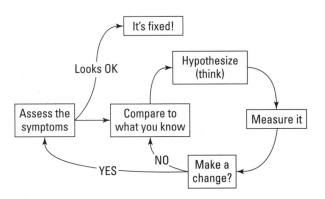

Figure 13-1: The troubleshooting process is a cycle of observation, thinking, measuring, and observing once again.

Keep this process in mind as you troubleshoot. Particularly with difficult problems, having the rules of the game in mind can help a lot!

Organize your thoughts

You wouldn't play a complicated game with all the game pieces and cards in a giant heap (would you?) — and it's the same with troubleshooting. As you go around and around, it's easy to lose your place. Unless you organize your thoughts, that is.

Keep a running log

Staying organized requires nothing more special than a pen and paper, as you can see in Figure 13-2. As you work through the problem, record your observations, your hypotheses, your measurements, thoughts that pop into your head, and anything else you notice or learn. That's what the log is — a record of the events that occur during the troubleshooting process.

Figure 13-2 shows an example of a troubleshooting log on the buffer amplifier circuit in Chapter 6. Let's say you built the buffer amplifier and it was working fine. Then one evening a thunderstorm blew through — and the next morning your buffer wasn't working any more. Uh-oh, it's time for a troubleshooting session!

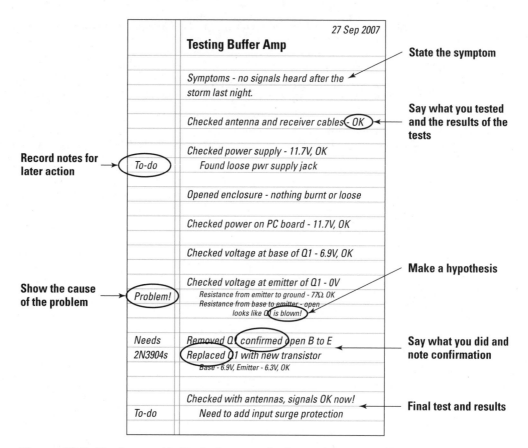

Figure 13-2: Keeping a troubleshooting log on notebook paper is an easy and inexpensive way to track problems and successes during troubleshooting.

Start with a clear label and date so if you have to set aside the project for a while, you'll know which notes apply to which parts and when they were taken. Then describe the symptoms. In this case, the amplifier didn't work after the storm.

As you take each step, write down what you're doing, the results of any measurements, and whether the measurements were normal or not. As you proceed, you'll probably notice other things that might need doing later (such as fix that worn power supply cable and buy some more 2N3904 transistors). Note those things in the margins.

At some point, you'll probably find a probable or reasonable cause for the problem. In this case, no voltage at the transistor's emitter, indicating that the transistor itself may be bad. Note what makes you think it's a problem ("Resistance from base to emitter — open!"). If you make a hypothesis ("Looks like Q1 is blown"), write that down.

Whatever you decide to change — an adjustment, a component, a connector — note what you did, and whether any additional measurements were made. In this case, the suspected transistor was tested — and found to indeed have an open-circuited base-to-emitter junction. It was then replaced.

Note the results of the change. If this fixes the problem — or causes other measurements (such as circuit voltages) to return to normal — note what seems to have fixed the problem. You might want to make more detailed measurements later.

A troubleshooting log is even more important during the debugging process when you're trying to get something to work the *first* time! You'll be making lots of measurements and changes. Without a log, it's easy to get lost and waste a lot of time!

Record what you learned

After you've completed the troubleshooting, get out your notebook (it should never be far away from your workspace!) and write down what you've discovered. Take a photo or make a sketch if the information needs a visual description to be clear. Remember that you might not need that information for a long time — well after the details fade from memory.

Then go meet with your friends and share your latest story with them. Everyone loves a good troubleshooting tale. They'll pepper you with questions, tell a tale or two of their own, and the whole group will learn something. If the information is really valuable, consider giving a short talk or demonstration to your club or team. That's how we all learn!

Practice makes perfect

Although I don't recommend that you intentionally break equipment or damage your circuits, I do urge you to take the opportunity to attempt repairs when you can. The very best way to learn about any piece of equipment is to repair it! By working on equipment you didn't build or design, you'll learn to unravel many small problems in the course of the troubleshooting session.

If all your equipment is working just fine, why not go scrounge something from a friend or relative? Garage sales and thrift stores are also full of stuff that doesn't work perfectly and can be fixed. Sometimes a repaired item can be resold at your own garage sale!

Coming to your senses

Certainly, instruments such as multimeters and counters are required to make electrical measurements, but you also possess your regular senses to help you zero in on trouble spots.

- ✔ **Your eyes:** Learn to scan the outside and inside of equipment and listen to that little voice in your head saying, "That looks funny!" Hold things up to the light or look at them from different or unusual angles. Learn to spot discolorations from heat or corrosion. Learn what components and parts are supposed to look like.

- ✔ **Your nose:** One of the least mentioned, but most frequently used workshop senses is that of smell. Take note when something smells hot or there's an acrid smell around a piece of equipment. You'll soon learn to discriminate between the aroma of roasted resistor, toasted transistor, and cooked capacitor.

 ✔ **Your ears:** Learn to listen for unwanted noise, audio levels that are too high or low, subtle characteristics of the different types of distortion, and hums or buzzes where they shouldn't be. These are clues that augment the electrical measurements.

 ✔ **Your touch:** Even if you don't touch live circuitry (and you shouldn't!), touch will tell you when equipment or parts are too hot or not warm enough. Vibrations or wobbles are telling tips of imbalance or wear.

Pondering Power Problems

To paraphrase a popular saying, "If the power supply ain't happy, ain't nobody happy!" For electronic devices to work properly and reliably, the power source has to be happy. Ask experienced troubleshooters where they begin looking for the problem and chances are they'll tell you they check the power first.

Power problems can begin at the wall socket. Repairmen who make house calls will tell you that many times the problem of an appliance that "doesn't work" is just that the line cord or "wall wart" transformer (the small power modules that plug into the wall socket) was unplugged or just loose. AC outlets can be loose or defective; circuit breakers may have been tripped, as well. So if your device runs from AC line current, make sure it's plugged in securely — and that the socket is live — by using a tester (as in Figure 13-3). You might save yourself a lot of time!

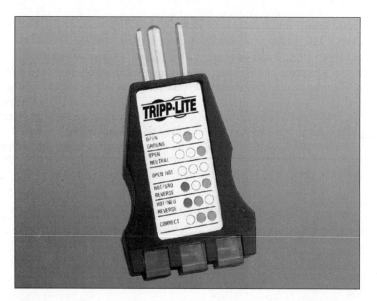

Figure 13-3: An AC socket tester quickly identifies a number of faults, including a "dead" socket.

Power troubles can masquerade as other problems — hum, distortion, erratic operation, overheating, to name just a few. Whether the voltage is too high or too low, it's likely to have an effect on the circuits downstream. If you're chasing a gremlin that seems to be everywhere and nowhere at the same time, take a close look at the circuit power.

Fuses and breakers

Fuses and circuit breakers are supposed to interrupt power when there's an overload. It isn't always obvious, however, that they've tripped. If the overload is severe (say, from a short circuit), the fuse is blown — pretty obvious. If the overload is light and prolonged, however, the fuse may simply sag until the element breaks — and that might not be obvious at first look. Fuse elements have also been known to break from vibration or high temperatures; this often occurs inside the metal end cap of glass cartridge fuses where it's hard to see. Don't assume the fuse is good — measure its resistance.

Circuit breakers mounted on equipment usually have a small button that pops out of the housing when the breaker trips. To reset the breaker, you push the button back in. The button may be loose after tripping and not fully extended, so it's not always clear that the breaker is really tripped. Double-check by testing voltage into and out of the breaker.

A *GFCI (ground-fault circuit interrupter)* breaker trips if current between the hot and neutral wires is temporarily unbalanced. This is an indication of a shock hazard. GFCI breakers are installed in special sockets from which other sockets are wired. Even though the main breaker for the circuit may not be tripped, if the GFCI breaker has tripped the downstream sockets will not have power. GFCI outlets will often trip at current levels far below that of the main breaker and have been known to trip because they detect the signals of nearby transmitters.

Battery power

Batteries provide a steady source of DC, but not without their own set of special problems. (Battery types and capacities are discussed in Chapter 15.) The most common problem is a weak battery whose output voltage is fine — until current is drawn from it, at which time the voltage drops dramatically. You can see this effect in a flashlight using an incandescent bulb. With weak batteries, the bulb may burn brightly for an instant when turned on, then quickly fade to a dim light or go out entirely. When checking batteries, don't take their *open-circuit voltage* at face value. Test the voltage while drawing a few mA out of the battery with a 1 kΩ resistor, or use a battery tester, shown in Figure 13-4, that applies a load to the battery.

Battery holders have been known to make poor contact, leaving the device user scratching his head after inserting fresh batteries, to no avail. If a battery has corroded in the holder, the resulting damage to the holder contacts can create a non- or poorly-conducting layer that prevents the battery terminal from making contact. Holders that aren't too badly corroded can be repaired by using a brass or steel brush to clean the contact.

Figure 13-4: A battery tester applies a small load to the battery when measuring battery voltage, causing the voltage from weak batteries to drop, which identifies a marginal battery that would otherwise test as good.

Power troubleshooting guide

This part of the chapter is a general guide to help you organize your thoughts and attack power problems in a logical way. This process is presented in the form of a *flow chart* in which symbols represent actions, conditions, and decisions. Square boxes represent actions ("Test AC socket"). Diamonds represent decisions ("Is AC on?") and the conditions for the decision ("Yes" or "No") are shown on the arrowed lines indicating the path through the chart. (For more information on flow charts, read the Wikipedia entry at http://en.wikipedia.org/wiki/Flow_chart.)

Figure 13-5 shows a simple process of dealing with that most common of objects — an AC-powered device that won't turn on. This is good practice for more complicated troubleshooting and helps put you in that logical, "Just the facts, ma'am" frame of mind so critical to finding and repairing problems. The chart assumes that the device has an internal power supply that converts the AC to DC for the electronics — and that there is a power switch to turn the device on and off. If the device doesn't have fuses or circuit breakers, assume a YES for those steps and proceed.

The goal of the flow chart is to find any problems that may be present between the AC wall socket and the device's internal power supply. Every possible wiring configuration isn't covered, of course, but if you follow the general procedure, you're likely to find any problems that exist upstream of the internal power supply.

If you troubleshoot your way to the device's internal power supply, a new flow chart is in order. The most common *linear* power supply is usually a simple circuit consisting of a step-down transformer, a rectifier, a filter to smooth the rectified AC (pulses of DC current) into steady DC, and some voltage regulating circuitry. (An introductory description of power-supply circuits is online at www.williamson-labs.com.) Although a detailed troubleshooting plan for any linear supply is beyond the scope of this book, the flow chart in Figure 13-6 will provide some insight into whether your power supply is working the way it should. The flow chart assumes that AC input to the supply circuitry is all right (as determined by the flow chart in Figure 13-5) The downstream supplied circuitry may also overload the power supply or have its own problems, so be prepared to test it, too!

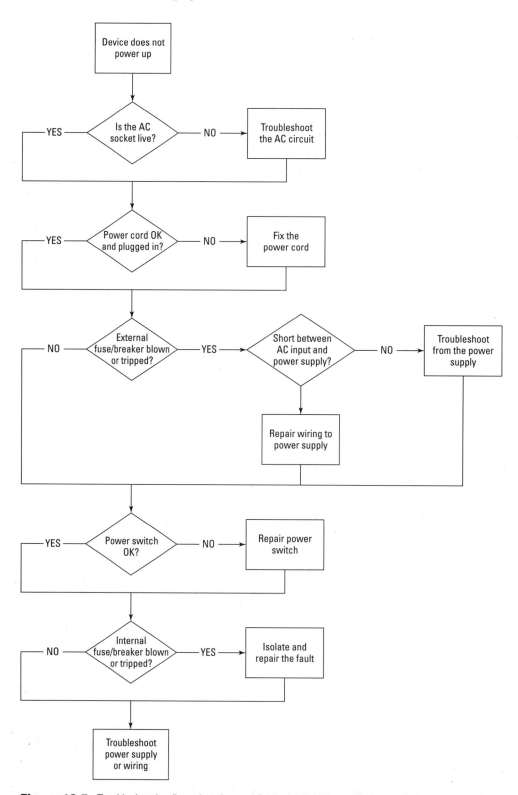

Figure 13-5: Troubleshooting flow chart for an AC-powered device that doesn't turn on.

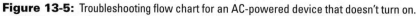

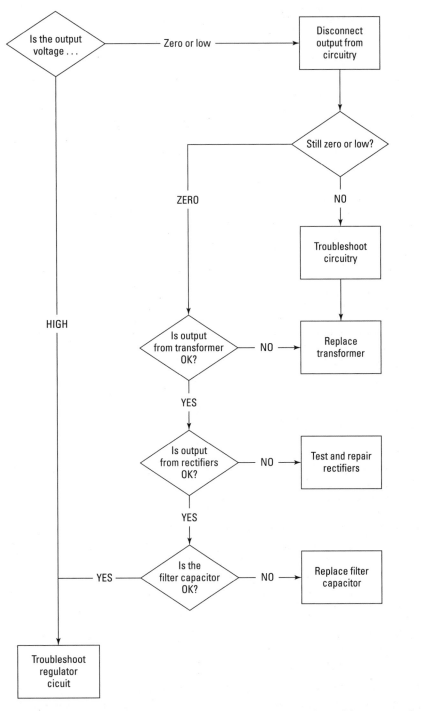

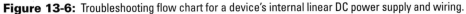

Figure 13-6: Troubleshooting flow chart for a device's internal linear DC power supply and wiring.

Common power supply problems

To help you look for problems, here's a list of common power supply failures and symptoms. Remember to follow the flow chart and don't jump to conclusions!

- ✔ **AC line fuse keeps blowing:** Assuming there's no short-circuit before the power supply, you may have a short-circuit from one of these causes:

  - A defective rectifier or filter capacitor acting like a short-circuit

  - A short-circuit in the circuitry supplied

  - The power supply has overvoltage protection, and that circuit may be drawing too much current to protect the supplied circuitry

- ✔ **Regulator circuit is hot and shuts down:** Usually a fault in the supplied circuitry that makes it draw too much current; output voltage will be low or zero.

- ✔ **Regulator transistors are blown:** If these are shorted, the output voltage will be too high; if they're open, the output voltage will be zero or very low; if they're damaged but not completely gone, the output voltage may be okay as long as nothing is connected to the power supply — but it'll be low when a load is applied.

- ✔ **Excessive ripple (hum or buzz in output audio or other signals):** Measure ripple on the power supply's output with a multimeter set to measure AC voltage; the usual cause is a defective filter capacitor at the rectifier output

- ✔ **Arcing (in high-voltage supplies, such as for televisions):** The usual cause is dust buildup — remove dust and lint with a vacuum cleaner's crevice attachment and an old (clean!) paintbrush.

Diagnosing Audio Problems

There are literally millions of different audio gadgets, yet most of the problems associated with them fall into a few categories with common causes. Figure 13-7 shows a flow chart that will help you through the following steps:

- ✔ Disconnect the equipment's input cables one at a time until the problem goes away. Replace the cable to check if it's just a bad cable. If the problem comes back, it's in the equipment attached to the other end of that cable.

- ✔ If the problem still exists after disconnecting all the cables, its cause is inside the equipment.

- ✔ Make sure the equipment's power source or internal power supply is working properly.

- ✔ In the case of hum or ground loop problems (covered in just a moment), try moving the equipment or changing the way the equipment is grounded. Reverse the AC line cord (if it's a two-prong cord) to see whether the problem changes.

These descriptions of problems and probable causes should help you analyze the symptoms and make your first hypothesis as discussed earlier.

Distortion

Distorted audio (at normal listening levels) reflects an inability of the circuit producing the signal to accurately follow the input signal. Assuming the circuitry itself isn't damaged or designed to introduce distortion in the first place, look for the following external causes:

- ✔ **Input signal too high:** Check the level of the audio signal at the input to the circuit or device. Distortion is a common result of connecting the input of high-gain circuits, such as microphone inputs, to line-level or speaker-level outputs.

- ✔ **Load is too heavy:** Trying to drive a heavy load, such as speakers, with a low-power device designed for headphones can overload the output circuit.

- ✔ **Power supply problems:** A power supply voltage that's too high or too low can upset the operation of amplifier circuits and cause distortion.

Hum and ripple

Hum is the contamination of signals with unwanted signals at the frequency of the power line, 60 Hz in North America. Hum is audible as a very low-pitched tone. It may be very loud or very soft, and is usually caused by electrical interference to the external connections to a piece of audio equipment, including microphones.

Hum can be caused by sensitive circuits being placed too close to a power transformer. The magnetic fields around the transformer can cause small signals to be coupled into the electronics, causing hum.

Ripple may seem like hum, but it sounds a little different. Ripple is present at *twice* the power line frequency, or 120 Hz for a 60Hz line frequency. It's the result of the rectification process that turns AC into DC in the power supply. (See the preceding section on power supply troubleshooting.)

Another source of hum is *ground loops* in which the ground connections between different pieces of equipment act as a pickup for magnetic fields, creating low-level hum in the audio circuits. To learn more about ground loops and how to eliminate them, read the tutorial at

www.angelfire.com/electronic/funwithtubes/Eliminating_Ground_Loops.html.

After you've decided whether the problem is hum or ripple, follow the flow chart of Figure 13-7 to further isolate the problem.

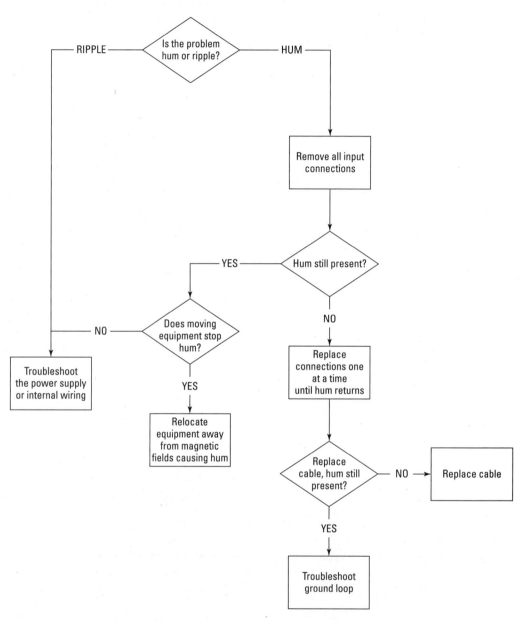

Figure 13-7: You can locate the problems that cause hums and buzzes by using this flow chart.

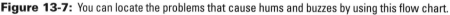

White and crackling noise

Two kinds of audio noise are the most common; hiss or *white noise* and a crackling kind of noise. Both are obnoxious if unwanted.

White noise gets its name because it's *colorless,* meaning equally strong at all frequencies. Another type of noise is *pink noise,* that decreases in amplitude with increasing frequency. Both are used for different purposes in audio systems and circuits.

Hiss at normal volumes generally indicates a defective audio circuit or device. You can quickly isolate the source of the hiss by disconnecting input connections until it goes away. If it doesn't go away, then it's being generated by the device you're listening to. Check all the controls to be sure you haven't accidentally caused high frequencies to be emphasized (such as turning off Dolby noise reduction on a Dolby-encoded tape playback deck) or left on some other amplifier circuit that has no input connection.

Crackling noise is usually caused by an intermittent or degraded contact somewhere in the system. Start by wiggling all input connections while listening to the noise. If one cable seems to cause the noise to start and stop, get louder, or disappear entirely, replace the cable and see whether the noise comes back. If wiggling cables doesn't affect the noise, disconnect the input connections until the noise goes away, then troubleshoot the noise source if you've found it.

After the source of the noise is identified as a specific piece of equipment, check all its knobs and switches to see if one affects the noise more strongly than the other. Rotary volume and level controls often become intermittent after frequent use or being left in one position for a very long time. Noise levels can often be reduced by rotating the control or switch back and forth to take advantage of the natural wiping action of the contacts. If the noise is not completely eliminated, spray the component's contacts (as shown in Figure 13-8) with a contact cleaner spray such as is available from RadioShack (www.radioshack.com). Do *not* use non-cleaning sprays such as WD-40 or household cleaners.

Figure 13-8: Cleaning a volume control by using contact cleaner.

Another type of crackling noise is caused by intermittent contacts at damaged or worn connectors. The contacts for jacks and plugs that are used frequently, such as for headphones or microphones can become worn or loose. The symptoms are usually crackling noises, hum, and buzz that are affected by wiggling the cable. Although the contacts in the connector can sometimes be bent or compressed to give a better connection, the noise almost always returns (usually at a completely inopportune moment). It's best to just replace the connector entirely, even if that is a pain.

Bad PC-board connections — particularly on headphone connections — can be caused by the mechanical stress of headphone connections cracking the solder joints of the headphone connector to the PC board. Unless the device is dropped with the headphones attached, the crack may be a hairline connection and difficult to see. Resolder all the connector contacts to the circuit board and use a magnifying glass to inspect the traces around the connector. Should any be found to be cracking (or already cracked), scrape away the solder mask over the trace and solder a short piece of fine wire across the crack to make the connection solid.

The techniques you learn to troubleshoot audio systems can also be applied to video equipment. Isolate the bad signal to just one input, check the cables, watch for sudden changes when moving equipment around, and be sure that the power supplies are clean.

Analyzing Analog Circuits

Unlike digital circuits that use specific voltages to represent ON and OFF signals, analog circuits (sometimes called *linear* circuits) operate in a continuous range of voltage and current. The most common *active* components in analog circuits are bipolar transistors, FETs, and op-amps.

This section assumes that you've already checked the power and signal paths and have isolated a problem to a specific *analog* circuit. In addition, you've used your senses and haven't found any burnt or overheated components. You'll have to discover the problem electrically. Here it's assumed that you have only a multimeter and that limits the tests you can do, but a surprising amount can be learned with some simple DC voltage measurements.

For guidance on making measurements, refer to Chapter 12.

After confirming that the power to the circuit is all okay, take a look at the schematic for the circuit. If you're lucky, it will have typical voltage readings on it inside small circles or squares and maybe even pointers to where the voltage is measured. If so, work your way from input to output, testing the voltages along the way. When you find voltages that vary by more than about 10 percent from the published voltages, take a closer look.

The circuit symbols show some basic examples of what voltages to expect when a transistor or op-amp is operating normally, amplifying a signal. If they are OFF (or otherwise electrically forced to some state in which they are disabled), you may measure all sorts of voltages. Nevertheless, by making quick checks of these voltages, you'll at least be able to tell which are operating and which need further inspection. Here's a quick list of the usual suspects:

- ✔ NPN transistors conduct current from collector (C) to emitter (E) when VBE is 0.6V or greater. This voltage does not change much with collector current level.

- ✔ PNP transistors conduct current from emitter (E) to collector (C) when VBE is less than –0.6V. The minus sign reflects the orientation of the measurement; from base to emitter. This voltage does not change much with collector current level.

- ✔ MOSFET transistors (enhancement mode) conduct current from drain (D) to source (S) when the voltage from gate (G) to source (S), VGS, is larger than the turn-on voltage, generally around 2V. Drain current varies with VGS.

- ✔ When an op-amp is acting as an amplifier, its inputs have essentially the same voltage. More than a few mV of voltage between these two terminals will cause the op-amp's output voltage to be driven to one of the power supply voltages.

Transistors are often operated as switches, where they operate in one of two states; *cutoff* or *saturated*. In cutoff (V_{BE} is close to 0), a transistor is not conducting collector current. In saturation (V_{CE} is less than 0.5V), the transistor is conducting as much current as the circuit will allow. These are normal states and may be intentional. In MOSFETs, the analogous states occur when V_{GS} is less than the turn-on voltage or when V_{DS} is less than 0.1V, respectively. Similarly, the output of an op-amp configured to act as a *comparator* will have its output forced to the level of the positive or negative power-supply voltage, depending on whether the voltage at the + or – input is greater.

Use Ohm's Law

Remember that you can use Ohm's Law (see Chapter 12) with multimeter measurements and resistor value markings to learn a lot about how your circuit is working. For example, if the voltage across a 75Ω resistor, R3, is 1.2V, you can calculate the current as 1.2 / 75 = 0.016 A = 16 mA. From your knowledge of the circuit's design, you can then decide whether that is appropriate.

Diagnosing Digital Circuits

In one particular way, digital circuits are easier to diagnose with a multimeter than analog circuits: All the signals are ON or OFF, corresponding to specific voltages. On the other hand, many such circuits depend on pulsed or constantly changing clocks — which puts them beyond the multimeter's ability to measure. In such cases, it's much better to have a logic probe (see Chapter 11) that translates its voltage measurements into displays that make sense for a digital circuit: ON, OFF, Active, Pulsing, and so on.

As in the previous section on analog circuits, this discussion assumes that you've already tested the power supply and isolated the problem by tracing it to the circuit you're about to test. Furthermore, I'm assuming that none of the ICs are hot or otherwise indicating physical damage.

Start by confirming proper power-supply voltage on each and every IC in the circuit. Identify and test any control signals that might cause the circuit to pause or hold. These are usually shown as *enabling* (often labeled EN) or *disabling* inputs. Satisfy yourself that these signals are in the proper state for the circuit to work.

If the circuit has one or more *clock* signals (streams of pulses), be sure they are active and present at every input to which the schematic shows they're connected.

Each IC probably has one or more *control* or *configuration* inputs that cause it to act in specific ways. For example, many digital ICs have inputs that control or configure the IC to act in a certain way. If any of these are not in the required state (0 or 1, corresponding to Low or High logic voltages), the IC will not function as required.

After you've confirmed all the ICs are configured properly, trace the signals through the circuit. Look for discrepancies between how the IC is configured and what happens to the various input and output signals.

Chapter 14

Maintaining Your Cool (Stuff)

· ·

Topics in this chapter

▶ Taking care of tools and testing instruments

▶ Maintaining a winning workspace

▶ Maintaining electronic equipment

▶ Keeping a schedule

· ·

Don't forget that you have invested time, effort, and hard-earned dollars in circuits, knowledge, and equipment! Those investments are like your car or home; you can preserve their value in the world of circuitbuilding by performing regular maintenance. If that conjures up images of greasy overalls and giant monkey wrenches, forget it! Just by planning ahead and organizing the job, you can ensure that the kind of maintenance described in this chapter can be performed in minutes.

Taking Care of Tools and Test Instruments

The most important stuff to maintain is your growing collection of tools and test instruments. After you've been working with electronics for a few years, you'll have several hundred dollars-worth of tools and parts accumulated, so it's important to take care of them. Besides, after you've grown used to a specific tool, it becomes part of your extended family!

"Clean and dry" is the mantra for tools. Electronic tools don't require frequent sharpening or oiling the way metalworking and woodworking tools do. But they *will* rust or corrode if allowed to get damp or wet! If you take tools outside overnight or work in the rain, be sure to wipe down your tools before putting them away. If you can, put them under a warm lamp (a desk or bench lamp will do) for an hour or so to let the heat of the light bulb evaporate the water or condensation.

Keeping track of loaned-out tools or books can be a problem, so make a note of every loan — in your permanent lab notebook. You're much less likely to misplace a tool that way; random slips of paper tend to . . . well . . . slip away.

Tool turtles

A turtle's home goes everywhere with the turtle. Your tools can go with you if they are kept in a good toolbox. Even though you may have a workspace set up at home, there will be plenty of times when you need to take a set of tools somewhere else; to work in a car, on a trip, at a field site or activity. If you keep the most-used tools in a toolkit like the one shown in the figure; it's easy to grab and go! It also keeps your tools organized so you can always find the one you want and tell if you have them all when packing up.

You can spend as much as you want on a toolbox — there are some mighty expensive cases out there. The biggest bargain for tools, however, is the common tackle box! Almost any time of the year, you can find inexpensive tackle boxes in the sporting goods sections of stores and they're on sale at the beginning of the outdoor season! Don't forget to look in the thrift and second-hand stores, too!

On the right in the figure is a great use for tackle boxes — holding and organizing RF cable connectors and adapters. Not only does it hold adapters, but a soldering gun, solder, crimping tools, and a few other handy items. Wherever the job, home or away, the tackle box holds everything the job requires. No more rooting through boxes and drawers trying to assemble a collection to take along.

With tackle boxes so inexpensive, why not have one for each type of job? One for computer network jobs, one for audio, one for RF, one for wire-wrapping, and so on. For a few dollars, you can make big progress in taming your unruly collection of "stuff."

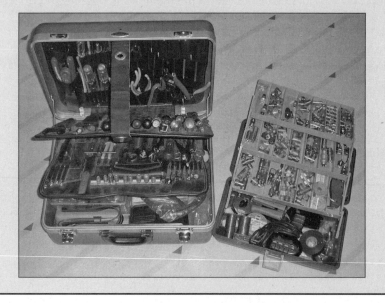

Test equipment is easy to maintain. Just keep it clean and don't let it get knocked around. From time to time check the *calibration* of the equipment, especially if you think the equipment may have been subjected to an overload or mechanical abuse. That doesn't mean you need to send it to a special lab, just that you need to confirm that it works properly and is giving you reasonably accurate readings. That's fine for a home lab.

Build your own voltage calibrator

It's very useful to have a known voltage handy to check meters and other test equipment. The simple circuit shown in the figure provides a stable DC voltage at the output. The 1N4733A Zener diode maintains a constant 5.1 V from its anode to cathode when current flows through it in that direction. (For more information about Zener diodes, check out http://en.wikipedia.org/wiki/Zener_diode.) A 9V battery is used as the power source and the 390 Ω resistor limits current through the Zener to a safe value. The voltage across the Zener diode is reduced by the 5 kΩ adjustable potentiometer. The 0.1 µF capacitor filters out any noise.

You can build this circuit using any of the techniques in Chapter 6. For output connections, banana jacks are recommended, but you can use any convenient connector — terminal strips, a phono jack, even machine screws, whatever provides a convenient connection.

After the circuit is constructed, connect a 9V battery to the unit and turn it on. Confirm that there is 5.1 V across the potentiometer element. (If there is only 0.6 V or so, you have the Zener wired backwards.) Adjust the potentiometer until you get the desired output voltage, which should be between 0.5 and 4 V. 1.50 V is a useful value because it's close to full scale on many multimeters. If possible, perform the adjustment using a calibrated multimeter, such as a friend might have or have access to.

Here's a parts list for building a voltage calibrator:

Enclosure

9V battery clip and 9V battery

Toggle switch

390 Ω, ¼-watt resistor

5k Ω potentiometer

0.1 µF ceramic capacitor

Red banana jack

Black banana jack

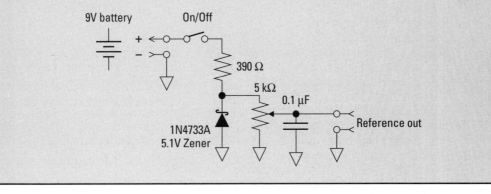

An easy way to have confidence that your test equipment is working is to use the different pieces of equipment to test each other. For example, if you have a pair of multimeters, use them to measure the same resistor — and watch for slow changes in the readings from either of the meters. You can also design the voltage calibrator in the "Build Your Own Voltage Calibrator" sidebar to have a known voltage to test. If you use the calibrator to test all your meters over time, you'll have a good idea of when one of them isn't working or has drifted out of alignment.

If you have a friend with calibrated instruments or can make friends with a local business or technical school, have them measure precisely the values of three components for you; a resistor around 1 kΩ, a capacitor of a few hundred pF, and a small inductor of 1 µH or so. Record the exact measured value of these components and keep them together as a set with the measured values. These are your new lab standards! Ham radio and robotics clubs are great ways to meet technical professionals with access to calibrated lab equipment.

If a piece of test equipment has special probes, connectors, or adapters, keep them and the instrument in one cardboard box or in the carrying case between work sessions.

Maintaining a Winning Workspace

The most important thing you can do to maintain your workspace is to keep it organized and clean. Although some folks don't mind a cluttered workbench, they also don't seem to mind the regular lost parts, misplaced tools, and electrical mishaps that come from loose wires and random bits of this and that. All it takes to keep things tidy is a whisk broom to brush debris into a trash can. (The occasional vacuum-cleaning won't hurt, either.)

When you *do* vacuum the workspace, use a new bag or empty the old one *first*. Then, when you suck up that irreplaceable part you lost last week, you'll be able to retrieve it from the bag.

Before beginning a building session and immediately afterward, take a few minutes to organize the workspace. Try to keep tools and test instruments in a consistent location. When you're working, it's frustrating to have to stop for a tool hunt or to locate a crucial test instrument.

Prowl the kitchen and housewares aisles of your local department store for specials and bargains on organizing trays, storage boxes, and racks. These cost much less than items made specifically for electronics — and they're made much the same.

A rarely-mentioned benefit of having an organized workspace is that it reduces disruptions to your train of thought. When you're trying to build or troubleshoot a circuit, it's important to keep track of what you're doing. If you have your tools and components where you can grab them without thinking about it, your brain stays focused on the job at hand.

Clip leads and test cables can be kept organized with a test lead holder, such as the Pomona Electronics 1508 (www.pomonaelectronics.com). Make your own if you prefer!

Another source of endless frustration is not having the right part when you need it. As you build more circuits, you'll notice that you use certain values and types of components and types of mechanical parts more frequently. You can

✔ Buy quantities of those parts and build up a small stock.

✔ Watch for sales on parts, especially parts assortments.

✔ Build up a network of friends who also build circuits; you can trade with them.

✔ Keep a small notebook with your parts and note when you're low on or have used the last of a part. Take it along when you go parts shopping. It will help you get the parts you need and also cuts down on impulsive buying decisions.

Maintaining Electronic Equipment

Regular cleaning is about all the maintenance that most electronic equipment needs. When you clean up the workspace, do the same for the equipment. Wipe it down with a damp (*not* wet) cloth. Household cleaners shouldn't be used; they usually leave a residue. Never spray liquids onto electronic equipment! Moisten a cloth or rag first, and then wipe.

One of the best-kept secrets seems to be that lighter fluid is a great solvent for grease, tar, adhesive residue, and all manner of organic junk that gets on tools and equipment. Use a tissue or cotton swab to minimize the amount you use as it evaporates very quickly. Also use in a well-ventilated area — and remember (oh, yeah) that lighter fluid is flammable; keep it away from hot wires or open flames. It also removes the oil on your skin, so wearing rubber gloves is a good idea. If you're using it on plastic, test it first on an inconspicuous place to avoid disfiguring your equipment.

As you clean the equipment, check the cables, power cords, and connectors. If they are loose or frayed or cracked, replace or repair them. This is also a good time to be sure you can locate any special cables or adapters that go with the equipment; avoid the dreaded phrase, "I just *saw* that thing."

If you don't use the equipment regularly, turn it on and let it warm up. Check its functions with your other test equipment. Be sure it's working the way it should. Run switches and controls through their ranges to prevent oxide buildup or dead spots.

Portable and mobile electronics

Electronic stuff that spends most or all of its life outside gets knocked around a lot more than the workbench-bound variety. Because of the many bumps and bruises, you need to pay closer attention to equipment used in these environments. (See also Chapter 16 on installing equipment in vehicles.)

✔ **Cables:** These get bent, yanked, tripped over, mashed; they take all sorts of abuse. The connectors they attach to also suffer. Give these a close inspection from time to time and look for crimps and solder joints coming loose. Connector pins and sleeves can get bent or pushed back into the connector. Cable jackets get cut or scuffed (repair them with quality electrical tape). If coaxial cable gets squashed (say, in a car door), it can develop short circuits or stop acting as a good pipe for radio signals — replace the cable if it looks damaged. Tighten loose connectors on equipment and check the connections inside; the vibration or twisting may have affected them, too.

✔ **Dirt:** Give the equipment a good cleaning, including inside the enclosure or cabinet. Bugs and dirt will find a way to get inside. This is the time to remove dirt and residue on the outside of the enclosure — before it hardens, dries on, or stains the finish.

✔ **Heat:** Anything mounted in or routed through an engine compartment or near a heater needs regular inspection. Heat causes many plastics to become brittle or deform. Corrosion or contamination happens faster at elevated temperatures, too. Batteries and sensors often have shorter lives when operated at high temperatures.

✔ **Carrying cases and handles:** A close visual and mechanical inspection might turn up cracked handles, loose mounts, broken hinges, and so on. Fix them now or put them on your to-do list (near the top)!

✔ **Accessories:** Did you get home with all your coaxial adapters? How about that mono-to-stereo miniature phone plug converter? Did you leave the DC power cable in your friend's car? This is the time to start the retrieval process, not the night before your next trip!

Equipment overboard!

It will happen sooner or later — a valuable piece of gear falls in the water, gets left out in the rain, or gets doused by something spilling on it. Before you start browsing the 'Net for its replacement, try the following procedure. It has revived a lot of equipment otherwise headed for the landfill.

✔ Take the equipment's cover or enclosure off and hose EVERYTHING down with fresh, clean water. Bathtubs or laundry sinks are good for this purpose.

✔ If the equipment was in the water more than a few minutes — or if it has dried out since being dunked — soak it in fresh, clean water for at least a couple of hours. Then hose it down again.

✔ If the equipment can withstand being spray-cleaned, you can run it through the dishwasher once or twice before proceeding further.

✔ If the equipment was in dirty water with mud or silt or organic material, use a paintbrush and toothbrush to remove all the gunk under a stream of — you guessed it — fresh, clean water.

✔ Take all internal cable connectors apart and clean them with a toothbrush or cloth. Label or sketch any connections that won't be obvious when it's time to put them back together.

✔ Shake all the water out of the equipment and then place it in an open oven set on Warm. Place a meat thermometer next to the equipment and don't let the temperature go over 150 degrees. Keep the equipment in the oven for a couple of hours.

✔ Take the equipment out of the oven and let it cool.

✔ Give it a very close inspection and repair any visible damage. Reconnect all the connectors.

✔ Cross your fingers and power it up!

Steam-powered equipment

You may not think of monitors and TVs as vacuum-tube equipment, but they are! The CRTs (Cathode Ray Tubes) in these big boxes are every bit as much vacuum tubes as their test-tube-sized relatives in your grandparents' antique radio. Turn off the room lights and look inside at the very back of the CRT neck. Inside you'll see the glowing signature of the tube — a filament that heats a cylindrical cathode. Electrons are boiled off the cathode into the surrounding vacuum, where they are snatched up by electric fields to be whisked past controlling grids and through steering magnetic fields to the front of the CRT. The inside of the CRT is coated with special phosphors that glow from absorbing the energy of the electrons. Yes, you are using real, live vacuum tubes, the "steam power" of electronics!

Mostly these boxes require little maintenance — until one day you notice that the picture jumps around or breaks up from time to time. You may also hear a little snap or crackle from the back of the monitor or TV. What's happening is that dust is attracted to the CRT by high voltages on the tube's electrodes. As it builds up, the coating of dust creates an unwanted conductive path that eventually allows the high voltages to occasionally arc to ground. The arc disrupts the voltages that control the electrons in tube — so the picture is also disrupted for a moment. If you don't get rid of that dust, the arcs will get stronger and last longer until the picture is unwatchable. The high-voltage power supply can also be damaged, resulting in a repair bill . . . or a new monitor or TV!

The solution is to break out the vacuum cleaner. (How about that? Using a vacuum cleaner to repair a vacuum tube!) Find a paintbrush about 1" to 1½" wide (used with relatively clean bristles is fine) and some rags or soft cloths you don't mind getting dirty. Place the afflicted display on your workspace face-down.

Before proceeding, consider that you're about to work on equipment that operates with some seriously high voltages. First and foremost, remove that power cord! Don't use metal implements (use the plastic attachments) and don't touch exposed components. Don't remove the high-voltage lead that goes to the side of the CRT. Once you've completed your cleaning, don't operate the display with the case open. OK, now get on with the job!

Look for the four to six screws that hold the plastic back on the case and remove them. You'll probably see a lot of . . . DUST . . . stuck to the surface of the CRT, the thick cable going to what looks like a suction cup on one side of the CRT, and a complicated, vaguely sinister-looking assembly on the PC board (a high-voltage power supply) to which the cable is attached.

Your mission is to remove that dust! Pretend you're an electronic dental assistant; use the vacuum's crevice cleaner to suck up the dust as you loosen it with the paintbrush. Be careful around the fragile back of the CRT where the wires are attached. Don't move — or even jiggle — the wire coils around the neck of the tube or your pictures will be forever distorted. Gently remove all the dust you can, wiping down the surface of the tube, the cable, the sinister assembly, and the inside of the cabinet with a cloth. It will not be necessary to remove the PC board or the tube — just clean them. Reassemble the case and it's a pretty good bet that the "cracklies" will be gone. If they aren't, the power supply may be in need of repair — and you'll have to have a repair service take care of that. You can perform this same maintenance on any tube equipment, including musical-instrument amplifiers and antique radios.

Although dust is one enemy of tube-based devices, the other mortal enemy is heat. That's why the display enclosures have holes for the dust to get in — so heat can get out! Don't block these holes with dust covers or newspapers or pizza boxes. The resulting heat build-up causes premature aging of the electronic components. Let plenty of air circulate. And vacuum every once in a while, will ya?

Electronics with moving parts

A gear that has motors, wheels, solenoids, and so on has special maintenance needs. These gizmos are usually lubricated — and the grease or oil can also get on the electronics and into the connectors. Cables can easily get chafed or pinched. These devices are also usually in places dirtier than an electronics workspace. All these make maintenance a little more involved than for purely electronic devices:

- ✔ **Cleaning the equipment is probably the first and most needed step.** Whether the outside surfaces get attention or not, the internal electronics needs to be clean. Open the equipment and remove any buildup of dirt and debris. If grease and oil are getting on to the electronics, remove the gunk. This is the time to see whether shields are needed to protect the electronics.

- ✔ **Connectors and sensors often get coated with dirt or dust.** Connectors can become contaminated with lubricants or solvents that get into the pins and sockets. Remove connectors and inspect the inside surfaces — they should be clean, and the metal surfaces bright and shiny. Sensors should be clean, and mechanical switches free of obstructions and debris.

- ✔ **Inspect for mechanical problems like cables getting pinched or pulled.** Wherever cables or wire bundles go through enclosures or walls, be sure that they are protected by grommets or sleeves in good condition.

Maintaining yourself

Yes, you! You need to maintain your skills and your knowledge so don't leave yourself out of the maintenance rotation. It's easy for your technical groove to turn into a technology rut. You will have a lot more fun building circuits if you stay fresh by learning new skills and keeping your existing skills sharp.

Classes in technical specialties are often taught at local community colleges, libraries, and technical schools. They may not be offered regularly, so get on an e-mail list or put the appropriate Web sites on your list for regular visits. These are usually inexpensive or even free, so don't miss out on the chance. The instructor may also be a great resource for information and references beyond the class.

Online tutorials abound, so take some time now and then to review — or expand your horizons with something new. Appendix A has a list of resources that includes numerous online education sites.

Licensing and certification can make a difference at work or in your professional life. Have you ever wanted to get a ham-radio or commercial radiotelephone license? Maybe a network or welding certification would be good? You can also build up recognized credits by taking courses at the local community college or technical school.

Trade and craft shows don't cost much to attend but they provide an enormous opportunity to meet others who have the skills you need — and who came to meet and talk to people like you! Demonstrations are often performed that may give you exactly the ideas you need for a project or provide the spark for a new one!

Clubs of like-minded circuitbuilding enthusiasts are wonderful places to learn. You can also share what you've learned with others. You'll learn about the best places to get parts, techniques that do and don't work, and make friends, too.

Keeping on Schedule

None of this will happen unless you decide to make it happen! In this busy world where we have so many demands on our time, waiting for a free day and *then* deciding to do some maintenance makes the maintenance pretty unlikely. The solution is to put maintenance on your calendar, just as if it were a sporting event or a birthday. It's much more likely to happen if you plan for it in advance.

Take a sheet of paper and make three columns labeled "Yearly," "Quarterly," and "Monthly." Then sit down at your workspace and look around. Everything that needs maintenance should go in one of those columns. For openers . . .

✔ Don't forget batteries! Regular maintenance is a great time to replace batteries in flashlights, smoke alarms, radios, and so forth. (This is also a good opportunity to gradually change the batteries you use in your electronics to use rechargeable batteries so you won't fill up the local dump quite so fast.)

✔ What goes in the yearly column? Test equipment. Cleaning out your toolbox. Infrequently changed batteries. Recertification, subscription, and license renewals.

✔ How about quarterly? Clean up your workspace. Check for upcoming classes or shows in your area. Go through your loaned-out list (or return tools you've borrowed!).

✔ Does anything need to be done monthly? The replacement-parts shopping trip. Inspection of your mobile radio equipment and antennas. Making sure all your tools are where they're supposed to be.

These are only suggestions, but use them to get started on your own list. The old proverb says that the journey is halfway complete after taking the first step. It certainly couldn't be more true than for maintenance!

Chapter 15

Getting a Charge Out of Batteries

- -

Topics in this chapter

▶ Types of batteries

▶ How batteries work

▶ Battery energy capacity

▶ Choosing between disposable and rechargeable batteries

▶ Charging and discharging batteries

▶ Battery safety

- -

For electronics, portability usually requires battery power — or a very, very long power cord. Even solar-power aficionados rely on batteries to store excess energy through the day. Batteries are everywhere and in every sort of device, and it seems there's a new type of battery every month! This chapter reviews the different battery types so you can decide what battery is right for the job.

A Bunch of Battery Basics

All the jargon associated with batteries makes a whole lot more sense if you know a little about what makes a battery go. In the simplest terms, a *battery* is a device that produces electricity through a chemical reaction. The interaction between self-contained chemicals is what makes batteries work; it's unnecessary to tap into another power source (such as a wall socket) when you use a battery. You increase the versatility and portability of your electronic devices when you use batteries — especially long-lasting batteries. You use several items every day that have batteries — your car, you laptop computer, your mobile phone, or your digital camera. Although you don't have to know chemistry to understand why batteries are important, you should know a little something about how (and how well) they work because you can waste a lot of cash on inefficient batteries.

How batteries work

A chemical reaction can generate electricity because chemicals hold onto their *electrons* (the smallest particle of electric charge) with different strengths. The strength is called *electropotential*. When two chemicals with different electropotentials are brought into contact, electrons transfer to the chemical that wants the electrons the most. You can't just mix the chemicals together; if you do, the electrons move without going through a *circuit*. And of course, you need a circuit if you want to make the electricity do some useful work (such as running an electronic gadget!).

If two chemicals are kept from mixing because they're separated by some kind of barrier, they quickly build up an imbalance of electrons as the greedy chemical collects them. The resulting imbalance between one side and the other creates the voltage difference you measure between the terminals of the battery. Every battery has two terminals, positive (+), and negative (–). If you connect a circuit from one terminal to the other, the voltage pushes the electrons through the circuit, supplying energy to run electronics.

The voltage between the terminals is called the *characteristic voltage* of the battery. The characteristic voltage depends on the type of chemicals used. When you go to the store and see alkaline, Nickel-Cadmium (Ni-Cad), or Lithium-Ion (Li-Ion) batteries, they all have a slightly different voltage. Battery manufacturers have come up with a number of chemical combinations that produce voltage.

The terms *dry-cell* and *wet-cell* battery refer to the makeup of the *electrolyte*, the barrier that separates the chemicals. Most types of batteries are a little bit wet, or at least damp. A dry-cell battery's electrolyte is a soggy paste, for example. A wet-cell battery, such as the one in your car, actually has a completely liquid sulfuric acid electrolyte (danger!). Gel-cells have the same chemicals in them as your car battery, but instead of liquid sulfuric acid, the electrolyte has been made into a gel to prevent spilling.

You can find a much more complete description of battery operation at http://science.howstuffworks.com/battery1.htm.

Standard voltage of alkaline or carbon-zinc batteries varies from 1.2 – 1.5V. The rectangular 9-volt (9V) battery is really made from six 1.5V batteries connected end-to-end so their voltages add up to 9V. *Lead-acid* batteries, such as those used in a car or a *gel-cell,* provide 2V per cell. Putting six of them together end-to-end makes a 12V battery, which is standard for automotive use. Batteries made with lithium have a characteristic voltage of 3 – 3.3V. These are just a few examples.

Batteries come in different sizes and shapes because they have different jobs to do (see Figure 15-1).

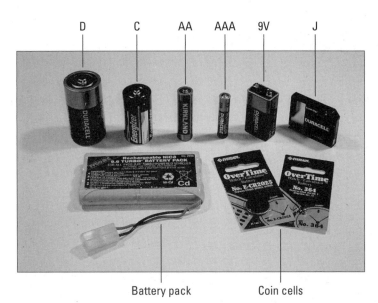

Figure 15-1: Different sizes and shapes of batteries are intended for different uses.

Ah . . . Introducing Amp-hours and Characteristic Voltage

Batteries have two primary ratings that you'll need to keep in mind to choose the right type for your circuit:

- ✔ **Characteristic voltage:** The relatively constant voltage between the two battery terminals. This measurement is based entirely upon the chemicals used to transfer electrons. (See the sidebar, elsewhere in this chapter "How Batteries Work," for more information on characteristic voltage.)

- ✔ **Energy capacity, measured in** *ampere-hours* **or** *Ah:* Because a battery's characteristic voltage is relatively constant, the only change between a little battery and a big battery of the same chemistry is how long the battery can supply current before it dies or requires recharging. If you multiply ampere-hours by the battery voltage, you can calculate *watt-hours* (watt-hours are the units of measurement used on most electric bills). The higher the capacity in Ah, the longer the battery will be able to supply current to your equipment while maintaining its rated voltage. More Ah equals more operating time.

If a battery can supply one ampere of current for an hour before exhausting its chemicals, it's rated at 1 Ah. A rechargeable Ni-Cad battery with a rating of 1200 mAh can supply 1200 mA (or 1.2 A) for an hour. The actual rating is more complicated than that, but you get the idea. Table 15-1 in the next section shows a number of common battery sizes, chemistries, and energy ratings.

Disposable Batteries versus Rechargeable Batteries

With some kinds of batteries, when the chemicals have transferred all their electrons, that's it; the battery is dead. Those are disposable or *non-rechargeable* batteries, also referred to as *primary* batteries. Other batteries are *rechargeable,* which means the chemical reaction can be run in both directions; one way creates electricity (called *discharging*) and the other way stores energy (called *charging*). These types of batteries are also known as *secondary* batteries.

Don't try to recharge a disposable battery because the chemicals just won't change back to the way they were. Attempting to recharge a disposable battery runs the risk of heating up the battery or causing major corrosion. They can even explode if the heating causes gas to build up inside the battery.

Disposable batteries and rechargeable batteries each have their own merits, which are discussed later in this chapter. If you're trying to decide between the various battery types, try regular, old-fashioned disposables and see how long they last. If you find that they die within a few days or weeks, you should really consider the more expensive rechargeable options.

Eventually, all batteries must be disposed of. Even rechargeable batteries eventually die. Be sure to keep extras around even if you rely primarily on rechargeable, long-life batteries.

Table 15-1	Common Battery Types and Ratings		
Battery Style	Chemistry	Voltage (with a Full Charge)	Energy Rating (Average)
AAA	Alkaline (disposable)	1.5V	1200 mAh
AA	Alkaline (disposable)	1.5V	2600 – 3200 mAh
	Zinc-Carbon (disposable)	1.5V	600 – 1200 mAh
	Nickel-Cadmium (Ni-Cad) (rechargeable)	1.2V	600 – 700 mA-hr
	Nickel-Metal Hydride (NiMH) (rechargeable)	1.2V	1500 – 2200 mAh
	Lithium-Ion (Li-Ion) (rechargeable)	3.3 – 3.6V	2100 – 2400 mAh
C	Alkaline (disposable)	1.5V	7.5 – 8.5 Ah
D	Alkaline (disposable)	1.5V	14 – 22 Ah
9V	Alkaline (disposable)	1.5V	580 mAh
	Nickel-Cadmium (Ni-Cad) (rechargeable)	7.2V	110 – 125 mAh
	Nickel-Metal Hydride (rechargeable)	7.2V	150 – 175 mAh
Coin cells	Lithium (disposable)	3 – 3.3V	25 – 1000 mAh

One factor in evaluating the quality of any battery (rechargeable and disposable alike) is the battery's *discharge curve*. In Figure 15-2, a graph represents how battery voltage changes as its energy is used up. A perfect battery would provide a constant voltage (represented in the figure as a horizontal line) until it's completely exhausted, and then drop vertically to 0 voltage. Real batteries try their best, but can't quite meet that standard. As they're discharged, their output voltage gradually drops to about 80 to 90 percent of the full-charge voltage. At some point, their output voltage starts to drop rapidly. If you think of the discharge curve as representing a battery-powered flashlight's brightness, you get an idea of how the battery performs as it's discharged.

In Figure 15-2, the two performance leaders are alkaline batteries (which are disposable) and NiMH batteries (which are rechargeable). The following sections rate all battery types, categorizing performance and cost effectiveness.

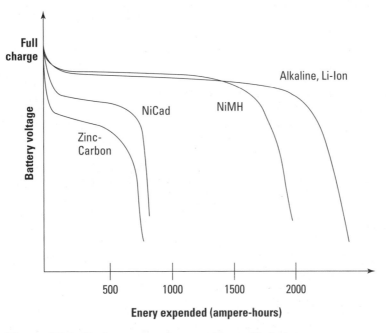

Figure 15-2: Discharge curves for several types of batteries.

Disposable batteries

You're probably most familiar with disposable batteries; after all, the alkaline battery has been immortalized by Energizer bunnies and Coppertops since the 1960s. Here are ratings for the two most common disposables around.

Zinc-carbon

Zinc-carbon batteries are a familiar type of disposable batteries. They're an obsolete type of chemistry and I don't recommend them. Although they function fine for a while, they have relatively short lives and often corrode. Between the rod and case of each zinc-carbon battery is a paste saturated with a weak acid — the very same acid that leaks out and ruins your flashlight when you leave discharged batteries inside. All in all, this is not a winning combination.

Zinc-carbon batteries show the most voltage drop as they're discharged. A flashlight using zinc-carbon batteries dims noticeably as the batteries grow weaker. Although this may be okay for a flashlight (although it's not optimal), most electronic devices need a constant voltage to work well.

If a battery's voltage drops too quickly, many devices will turn off before the battery's energy is completely used. If possible, use a more robust battery type.

Strength	Inexpensive
Weaknesses	Low energy rating and high voltage drop when discharged
	Energy capacity drops at low temperatures
	Not rechargeable and often corrodes when discharged

Alkaline

The alkaline battery is a disposable battery type whose chemistry is a big improvement over zinc-carbon. It uses an alkaline solution that's much less corrosive than the weak acid used in zinc-carbon batteries. Alkaline batteries also have a much higher *energy density* than zinc-carbon batteries, meaning that more energy is packed into the same volume. You can see evidence of this by reviewing the higher Ah ratings for alkaline batteries in Table 15-1.

The alkaline battery's discharge curve is much *flatter* than that of the zinc-carbon battery; the output voltage stays steady until a lot more of the stored energy is discharged. Because it contains more energy to begin with, the alkaline battery runs electronic devices for a much longer time than the zinc-carbon battery.

Strengths	Widely available with a high energy rating
	Does not corrode
Weaknesses	Expensive when used continuously
	Not rechargeable
	Heavier than alternatives

Disposable batteries contain chemicals that aren't exactly poisonous, but aren't good for the ground or water supply. Instead of throwing away your dead batteries, ask your local recycling center, electronics shop, or hardware store whether they can be recycled as described at the end of this chapter.

Rechargeable batteries

Rechargeable batteries generally give you the best of all possible worlds. They have long lives and are also rechargeable. Of course, not all rechargeable batteries are created equal, and cost is a consideration. The following sections sort through the issues and benefits of these batteries.

Nickel-cadmium (Ni-Cad)

The nickel-cadmium battery (Ni-Cad) was the first popular rechargeable battery type and is found in all sorts of rechargeable appliances, such as toothbrushes, drills, cordless and mobile phones, and model cars. Nifty as they are, there are better alternatives.

Cadmium is a toxic metal; although individual batteries pose no danger when used in electronic devices, after the battery finally dies you're faced with disposal problems in landfills. Cadmium can be leached into the ground soil, and this is not a good thing. (See the section in this chapter, "Safely disposing of batteries," for more information about proper battery disposal.)

Ni-Cad batteries (often pronounced "NYE-cad") have a decent energy rating, but they don't last as long as alkaline batteries. Their discharge curve is a little weak towards the end of charge, but they can be recharged many times at little expense. Ni-Cad batteries can also supply high bursts of current and that makes them a popular choice for cordless power tools.

Early generations of Ni-Cads had a *memory effect* that gradually reduced the battery's energy rating when recharged unless special conditioning was performed. The Ni-Cad batteries sold today are largely free of that problem. For low-power devices, Ni-Cad batteries are fine, but don't get too used to Ni-Cads; they're being phased out in favor of NiMH (nickel-metal hydride) batteries except where high discharge currents are needed, such as for power tools.

Strengths	Medium energy rating
	High discharge current
	Low cost
	Rechargeable
Weaknesses	Slight memory effect over many charge/discharge cycles is still possible, although this problem has been largely fixed
	Cadmium is toxic when disposed of improperly
	Lowest energy density of rechargeable battery types

Nickel-metal hydride (NiMH)

Nickel-metal hydride (NiMH) batteries were developed to better meet the higher power delivery demands of modern electronic gadgets like digital cameras, portable computers, and audio players. The cadmium used in Ni-Cad batteries is replaced with a compound of hydrogen and other metals — voilà, no more toxic metal. Additionally, overall battery performance is greatly improved.

NiMH is the first rechargeable battery type to exceed the alkaline's energy density. As a result, these batteries are rapidly displacing Ni-Cads. NiMH batteries hold their output voltage nearly constant for much longer than the Ni-Cad and are comparable to the performance of top-quality alkaline batteries.

NiMH battery chargers have become quite inexpensive and can recharge battery cells in just a couple of hours; this battery is highly recommended. They'll pay for themselves in no time!

Strengths	Excellent energy density
	Can be recharged quickly and for many cycles without "memory effect" power loss
	Less toxic waste to dispose of
Weakness	More expensive than alkalines

Lithium-Ion (Li-Ion)

The next step in battery development, Li-Ion batteries can be recharged many times over. These cells are often combined into battery packs and used to power laptop computers. They're just becoming available as separate cells.

Strengths	Excellent energy density
	Can be recharged quickly and for many cycles
Weakness	Most expensive of rechargeable types

Lead-acid

If you need a lot of energy, only a lead-acid battery will do the job. Although the lead compounds and strong acid electrolyte pose some toxicity problems, these batteries are inexpensive to manufacture and easy to recycle. You can see why this has been the battery of choice for most high-power uses for nearly 100 years.

Luckily for battery users, the hazards of the acidic electrolyte have been reduced with the creation of the gel-cell battery. Instead of a liquid, the acid is contained in a gel that prevents spillage with only a small loss in energy rating. The gel-cell is a safe and effective means of providing long-term energy storage.

Strengths	Good energy density and inexpensive
	Can be recharged many times
Weaknesses	Heavy
	Toxic materials and dangerous electrolyte require proper disposal

Exploring the World of Battery Packs

Many electronic gadgets come with or accept an assembly containing several individual batteries that are permanently wired together. These *battery packs* can be made to fit almost any shape required and in a variety of voltages.

If you crack open the case of a dead battery pack, you'll find several cells connected end to end in a series with welded tabs. It's generally not practical for the hobbyist to rebuild battery packs, but often they can be traded in or recycled.

Some tools and two-way radios are designed to accept battery packs of several different voltages, varying their output power as the voltage changes. You can choose a lightweight pack with a low energy rating or a heavier pack that gives more output.

For my own use, I try to have one or two of each type of battery pack.

Following Basic Battery Tips

Here, in no particular order, are some good ideas for getting the most out of your batteries:

- **Buy disposable batteries in bulk.** Buying alkaline batteries four at a time at the grocery store is convenient (and may seem less expensive at the point of purchase), but in the long run, it's incredibly expensive. Warehouse stores and online suppliers often have bulk packs of 50 or more batteries at a fraction of the price per battery if you were to buy them in small quantities.

- **Refrigerate spare batteries.** Because a battery is a chemical device and chemistry runs slower at lower temperatures, you'll prolong the life of your batteries by keeping them cold. Don't freeze them! The water in the electrolyte will expand and possibly crack the case, ruining the battery.

- **Always have a spare.** When buying batteries, always have extras for emergencies and periods of prolonged use. Don't be caught discharged!

  Because batteries allow you to take your gadget with you on the road, you should be extra diligent about bringing spares with you. Who knows when and where you'll find a plug for your charger?

- **Group rechargeable batteries in sets.** A set of batteries is only as strong as its weakest cell, so try to keep the set together as they age. Battery suppliers sell inexpensive plastic cases that hold a few batteries. Don't let one weak cell spoil the whole set.

- **Regularly condition rechargeable batteries to help keep them in top shape.** If you have several sets of batteries or battery packs, rotate them through a charge/discharge cycle on your charger every few months.

Adhering to the Rules of Battery Safety

Used properly, a battery is a safe and effective way of storing electrical energy. However, a battery stores a *lot* of energy and uses *concentrated chemicals* to do so. Remember that!

Letting the energy out too quickly can wreck the battery — and wreck whatever it's connected to. Exposing the chemicals or putting them under too much stress can result in danger to you and damage to the battery.

Charging and discharging batteries safely

To get the longest life and best performance out of your investment in rechargeable batteries, you should use a charger designed specifically for that battery type. For example, a Ni-Cad battery charger won't properly charge NiMH batteries. An improper charger may even damage the batteries or be damaged itself.

Good advice in any setting (not just batteries) is to do what the manufacturer tells you to do. The manufacturer wants you to have good results with its batteries and to be safe when using them (after all, that's the best way to get you to buy them again). Don't try to speed up battery charging with a high-power charger if the batteries aren't rated for it. Limit discharge current to within the battery's ratings. Use chargers on the batteries they're intended for. The reward for "pushing the envelope" is rarely worth the risks.

Here are some important things to remember:

- Set your smart battery charger to the correct setting. Smart chargers have a specific method of charging, called a *charging algorithm*. They're able to sense the battery's charge level and adjust the rate of charge so charge time and stress on the battery are minimized. The optimum algorithm varies with battery chemistry so Ni-Cad and NiMH batteries should be charged differently. If you have a smart charger, be sure it's set for the right battery type.

- Promptly remove charged batteries from your old-style battery charger. Old-style battery chargers just apply a certain amount of current to the battery until you remove it. Leaving a battery in such a charger can overheat and ruin a battery. Take care that batteries are removed promptly when charged.

- Follow the battery manufacturer's guidelines for charging and obtain a proper charger if necessary. A lead-acid battery charger must be able to switch to *trickle charge* or *float charge* automatically when full charge is reached. This keeps the battery at full charge without overcharging it.

- Let your batteries run down every so often. It does not damage Ni-Cad and NiMH batteries to be completely discharged. In fact, regular *100 percent discharges* help to restore the battery chemistry. Good chargers first discharge these batteries before charging them back up.

- Treat your lead-acid batteries right. *Deep-cycle lead-acid batteries,* such as those intended for RV and marine use, are also able to withstand an occasional deep discharge. (They work best, however, if never discharged to below 50 percent of capacity.) Regular automobile batteries, however, are not made for that type of use and will be damaged. Only completely discharge batteries that can handle it.

  The truly "charged up" reader will enjoy the Web page published by the Battery Tender company at `http://batterytender.com/battery_basics.php`. This site offers all you ever wanted to know about how to charge lead-acid batteries.

- **Don't stress out your batteries.** Batteries can also be stressed by discharging them too rapidly. Short-circuits and excessive loads can wreck internal connections and cause electrolytes to boil or vaporize. This type of use calls for special batteries with adequate *current surge* or *pulse* ratings.

Storing and handling batteries with care

Here's a short list of do's and don'ts to follow when you're handling batteries.

✔ **Don't freeze a battery.** The water in the electrolyte may expand, cracking the case.

✔ **Don't expose a battery to excessive temperatures.** High temperatures may result in too much pressure inside the case or damage to the chemicals.

✔ **Do keep batteries clean, dry, and off wet floors or shelving.** Keeping batteries from getting damp and dirty prevents slow discharge across the battery's surface.

✔ **Do use compounds that prevent terminal corrosion for any lead-acid battery that you store or use in an exposed location.**

✔ **Do take special care with wet cells, such as car batteries.** These batteries may leak small amounts of acidic electrolyte that can damage unprotected supports.

Batteries pack a real punch when it comes to delivering current. Even AA batteries can put out several amperes if short-circuited. This amount of energy is enough to make a battery hot enough to burn you or melt a small wire. Larger batteries can get hot enough to explode when they're short-circuited. A large lead-acid battery can start a fire or melt a tool when short-circuited with the potential for serious burns. Respect the energy. Protect battery terminals against accidental short circuits. For larger batteries, keep terminal protectors installed at all times.

Safely disposing of batteries

The chemicals that give batteries their great energy storage ability are also fairly toxic or corrosive. Don't throw old batteries in the trash where they wind up corroding in a landfill, leaching chemicals into the water supply — it may actually be illegal in your area! Hardware stores and battery stores will often recycle your old batteries, including alkalines, and often for free or a small fee. Your municipal or county government may also have a recycling program for batteries. Check www.battery recycling.com for more ideas.

Chapter 16

Electronics in Motion

There are many great examples of electronics to install in vehicles: radios, navigation devices, temperature and weather sensors, and audio gadgets. Vehicles have special conditions that builders and installers must take into account. This chapter gives you a look at some of the unique circumstances presented by mobile installations.

Learning About Mobile Installation

You have all the same installation considerations in a vehicle as you do at home — plus new issues such as vibration, temperature, and security to worry about. Just as when putting some new electronics to work at home, if you think things through first and take your time, you'll get the job done right without wasted effort. The key is to be informed!

Ask a dealer representing the manufacturer of your car for service bulletins or guides for installing mobile radios and stereos. Regardless of what you're actually installing, it will have much in common with this type of equipment. Because they sell so many cars and trucks to fleet owners, General Motors, Daimler-Chrysler, and Ford have Web sites or service bulletins discussing how to install radio equipment in their cars and trucks. These are excellent reading, whether you have one of their cars or not.

   ✔ **General Motors:** www.service.gm.com/techlineinfo/radio.html

   ✔ **DaimlerChrysler:** www.arrl.org/tis/info/pdf/INSTG01.pdf

   ✔ **Ford:** www.fordemc.com/docs/download/Mobile_Radio_Guide.pdf

Installing marine electronics

For those of you with boats, marine electronics installation is very similar to auto installation, with one exception — there's a lot more water! Seriously, the marine environment is not kind to electrical and electronic systems. Before installing your marine electronics, I suggest that you browse this section, and then proceed to a reference that deals specifically with marine electronics, such as one of these:

✔ *The Marine Electrical and Electronics Bible,* by John C. Payne (A & C Black)

✔ *The Powerboater's Guide to Electrical Systems: Maintenance, Troubleshooting, and Improvements,* by Edwin R. Sherman (International Marine/McGraw-Hill)

Both these books show you the special techniques and terminology that apply to shipboard electronics.

Enter your vehicle model and the search term *radio installation* or *audio installation* into an Internet search engine such as Google. For example, when installing a ham radio in my car, the first page of hits from a search for *Contour radio installation* contained www.gmrsweb.com/gmrsbille.html. This site shows how to install mobile radios in a car very much like mine!

You can find many more sites discussing how to install audio equipment in cars, discussing issues of obtaining power, routing cables, and mounting equipment. With some careful searching, you can turn up some information about your vehicle. You may be able to make use of some of the installation tools and materials available from audio installer sites, such as www.installer.com.

Understanding vehicle safety issues

Mechanical, electrical, and driving safety are all the utmost importance when installing equipment in a vehicle. Your safety and that of your passengers should be foremost in your mind at all times.

Mechanical safety means mounting equipment properly

Mechanical safety means attaching your equipment securely and placing the accessories so you minimize the possibility of endangering yourself or your passengers in an accident.

If your electronic devices (such as a radio, speaker, or GPS receiver) will be permanently mounted in a vehicle, resist the temptation to stuff them in the space next to your seat or let them sit on a seat or a dashboard. In an accident, the equipment will come loose and fly around inside the passenger compartment at high speed. Unsecured stuff inside a car causes a lot of unnecessary injuries. You should properly mount or secure the equipment, even if the use is just temporary. If the equipment requires an external antenna, it should be placed where a pedestrian (or you!) won't be easily poked by the antenna's tip. You'll put your eye out with that thing! (Yes, Mom . . .)

Electrical safety means maintaining your connections

Electrical safety primarily concerns proper power connections. Resist the urge to throw things together in a hurry to get on the road. (The author once nearly set his own vehicle on fire because he got sloppy with DC power behind the dashboard. Don't let this happen to you!) Take the time to identify the proper circuits, use automotive electrical hardware, protect your wiring from vibration and accidental shorts, and keep these big don'ts in mind:

✔ ***Never*** make an unfused connection to your vehicle's battery.

✔ ***Never*** work on wiring with the battery connected.

Driving safety means . . . well, driving safely

Don't let the use of the electronics impair your driving. If you view the controls and displays or make adjustments frequently, place them somewhere that minimizes the amount of time your eyes have to leave the road. Configure the equipment to minimize the amount of fiddling around you're required to do. If you find yourself distracted, pull over. Don't install equipment where it obscures the vehicle's controls or obstructs their operation. Just don't.

It's illegal in many states to install *anything* (electronic or not) in a location that obscures your vision of the road! Headphones are also illegal, and a growing number of states are placing restriction on the use of mobile phones while driving. All for very good reasons!

Tapping into vehicle power

For low-power devices, you can obtain DC current from a cigarette-lighter socket by using a lighter-plug cable or adapter. Late-model autos that may not have a lighter socket often have some kind of accessory power socket for powering laptop computers, mobile phones, and so on. These circuits are usually fused at 5 to 10 amps.

Resist the urge to put in a larger fuse and draw more current. The wiring for accessory circuits is not rated for the heavier load and may heat up. At the least, the voltage drop caused by higher-than-rated current may cause the electronics to operate improperly.

Here are some other important dos and don'ts:

✔ **Do disconnect the battery whenever you work on a vehicle's power system.** Disconnecting the battery eliminates the possibility of destructive and dangerous short circuits. All you have to do is remove one of the cables going to the battery, positive or negative.

✔ **Do be consistent and use red for all positive power wiring and black for ground.** Being consistent with the colors lets everyone else who works on your car know what the wires are used for. Use other colors for controls or audio connections.

Auto parts stores have all the parts you need for properly obtaining power from a car's (or RV's) power system. If you're a boater, you can find similar materials at a marine supply store, along with corrosion prevention compounds.

✔ **Don't use materials designed for indoor service in a vehicle or boat.**
Vehicle and marine service is a more severe environment. Vibration, temperature extremes, moisture, and exposure to oils and greases can quickly overwhelm light-duty home wiring materials. Figure 16-1 shows a few examples of power sources that are intended for use in a vehicle.

Figure 16-1: Vehicle power sources include lighter plugs and DC to AC inverters.

Using a fuse tap

It's often impractical to add a connection to the existing fuse block. Instead, use a *fuse tap,* such as those from Crowbar Electrical Parts (www.crowbarelectrical parts.com). Click the Fuse Holders button to find the adapters. These add a connection by using a spare fuse position.

You can find the hot side for the fuse tap by removing the fuse and then using your voltmeter to measure voltage at each terminal. The hot terminal is the one that has voltage with the fuse removed — and is usually the one closest to the middle of the fuse block. Your new circuit must be fused, so add an in-line fuse if the power cable doesn't include a fuse (see Chapter 7). This technique can be used for loads of up to 10 or 15 amps. Connect the ground lead directly to a nearby screw or bracket solidly connected to the vehicle's frame.

Some circuits are only energized when the ignition switch is turned on. If you want to use your gadget when the engine is turned off, remove the key and then test for voltage at the fuse block.

From an electrical standpoint, different sections of car and truck bodies (particularly those of trucks) are not always solidly connected together. Before assuming that connecting a negative power lead to a metal body part is the same as connecting it to the battery's negative terminal, use a multimeter to confirm the connection.

Setting up and maintaining battery connections

If your gadget (or collection of gadgets) is likely to draw more current than a fuse tap can handle, you should obtain power directly from the vehicle's battery. Connecting directly to the battery prevents stress to the vehicle wiring and results in a decrease in the voltage drop to your equipment. If you're installing manufactured equipment use the power cable provided. If the power cable isn't long enough to stretch from the battery, extend the cable with more wire of *at least* the same size as that in the original cable.

Of course, you have to find a hole in the vehicle's *firewall* between the engine and passenger compartments in order to get access to the battery. If you use an existing hole, it will likely have a rubber *grommet* or other protective material to keep the metal from chafing through the insulation. Use that grommet or protective covering for the vehicle wiring. If you have to drill a new hole, be sure to line it with an auto-motive grommet from the auto parts store. Don't depend on electrical tape over the wires to protect them.

In days gone by, it was often recommended that you run your power cable directly to the battery's terminals. For vehicles with the battery under a seat or in the trunk, this is still the right way. A better solution for today's cars is to run the positive wire to the battery's positive terminal. Run the negative wire to the point at which the bat-tery ground strap is attached to the vehicle — usually on a fender or the engine block. Both leads must be fused to protect the equipment in case of a short circuit! By connecting the negative lead to the vehicle instead of the battery terminal, you protect your equipment from becoming a ground path for other devices. Figure 16-2 illustrates three common methods for getting power from your vehicle; the cigarette lighter socket, a fuse tap, and connecting directly to the battery.

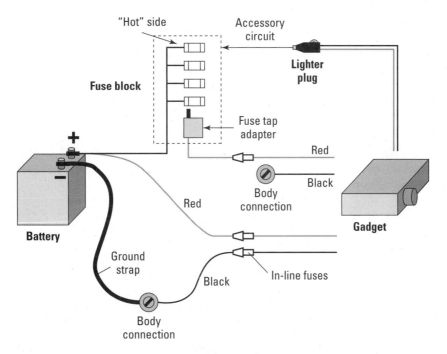

Figure 16-2: Three ways to make proper power connections.

Here are some additional installation tips:

- ✔ As you make your connections, secure the wires to supporting harnesses or brackets with nylon wire ties or electrical tape rated for automotive use. (At high and low temperatures, ordinary electrical tape adhesive will lose its grip.)

- ✔ Keep the wires away from moving parts and hot assemblies (such as exhaust system components or heater ducts).

- ✔ Protect the wires from sharp metal edges that might cut through the insulation aided by the vehicle's vibration.

- ✔ Don't position a connector or terminal so the wire is pulling on it — over time, vibration will work the wire loose.

Additional information is available in Web sites about installing vehicle audio systems and books such as *Car Audio For Dummies*.

Finding a home for electronics in your vehicle

In today's crowded cars, finding room for even more electronic gadgetry can be a challenge. How do you find a place so you can see it without it obstructing your ability to drive or getting in the way of your passengers? In general, you want to find a location from which the controls are easy to see and reachable without fumbling around. The device's cabinet must be mounted securely and have sufficient airflow to stay cool, if necessary. In order of desirability, here are several options:

- ✔ **Inside the dashboard:** This location is the best, but most of those spaces are usually filled with other electronics and controls, often making this a very difficult place to add equipment.

- ✔ **In the forward center console:** Look for *blind* panels that cover unused accessory-mounting bays. You can often remove ashtrays and storage compartments, as well. Not as difficult as the dashboard, but still crowded.

- ✔ **Under the dashboard:** This location can be hard to see (especially when you're driving), but it's an easy spot for installation purposes. Watch out for dangling cords or knee-gouging corners (especially if you have a manual transmission — ouch!). The device will also be difficult to operate if you're not in the correct seat.

- ✔ **On or to the side of the drive shaft hump:** In larger cars with center-seating capability, this is also an easy installation location, but often invades leg room space and makes viewing and operating inconvenient.

You should also consider anti-theft security. If you intend to remove your device when you're not in the vehicle, don't mount it so it will be difficult or time-consuming to disconnect. Removing it should take no more than 10 or 15 seconds. If you can't remove it, you may want to try for an installation where it can be hidden with a blank panel. Auto parts stores carry *slide mounts* so you can quickly remove a radio. These can be adapted to your gadget.

Mounting electronics under a seat or in the trunk can also work well if that is suitable for how it's to be operated. This has the added safety benefit of restraint in the event of an accident.

Consider building the equipment with a *remote control panel* at the end of a cable. This works well for switches, indicators, audio output, and power control. Because the remote panel is usually small and lightweight, it's easier to find a mounting location.

If the mounting location is under a seat, use some sturdy foam as protection from the bumps and bounces of the vehicle. A board or other cover protects your gear from accidental kicks by backseat passengers. Take care to route all cables so they aren't pinched by the seat sliding back and forth.

A unique puzzle comes along when you try to use a handheld device in the car. The best solution is to provide a sturdy pocket that hangs from the dashboard or console. Obtain a carrying case that's just a bit larger than needed and attach it to a convenient location. Cables can then be run to the pocket, where it's a simple matter to plug them in. A fabric or Velcro strap can secure the pocket while you're driving.

An old truth is that any cable cut to length is too short. Before cutting a wire or cable, make a test installation to be sure it's long enough. If you have to cut it before installing, leave plenty extra and then trim it. You can use the leftovers on other projects.

Chapter 17

Getting Rid of Interference and Noise

With electronic equipment seemingly everywhere and much of it wireless, is it any wonder that sometimes they interact in ways we don't expect? The result can be an occasional glitch, a bit of noise, or even complete disruption. This chapter will show you — a circuitbuilder and electronic-er — how to take the initiative and keep your equipment operating properly.

Dealing with Interference

Interference, either to or from your electronics, refers to unwanted signals or the effects caused by other signals. There are two types of interference: the kind you experience from external sources and the kind caused by your devices. We'll start with interference that your stuff receives externally — starting with radios and other RF receivers (some of them inadvertent) that you might have. Figure 17-1 shows several examples of interference-fighting devices and components as follows:

✔ **Telephone filters:** Installed in telephone lines.

✔ **Cables and wires:** Passed through or wound around ferrite cores.

✔ **High-pass filters:** Block lower-frequency signals while passing higher-frequency ones.

✔ **RF chokes:** Inductors (coils) used inside equipment to prevent interference to or from the equipment.

Telephone filter   Ferrite cores and beads   High-pass filter   RF chokes

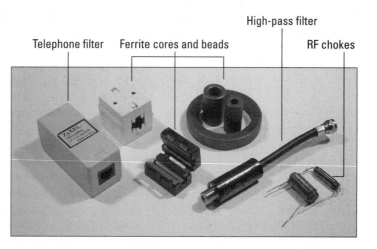

Figure 17-1: All these devices and components can be used to reduce or eliminate radio-frequency interference.

Received interference

The first thing to do when you are receiving interference from radio signals is to make sure your own radio or receiver isn't causing the problem. Modern radios have very sensitive receivers, but that makes them susceptible to interference or even causes them to generate false or *spurious* signals that seem like interference from another signal.

Overload is a common problem in urban and suburban areas. Sensitive receivers can only withstand so much received signal strength before their input circuits are overpowered by an unwanted signal. If you live near a strong broadcast or paging transmitter, you may experience overload. A strong signal from a nearby amateur radio station — operating completely legitimately — may overload nearby receivers. The interfering signal doesn't even need to be in the same frequency range that the receiver is tuned to. This is a common problem for stereos and televisions. When overload occurs, the receiver's output becomes distorted and erratic.

The resulting interference is not the fault of the transmitting station at all. The best solution is to reduce the strength of the interfering signal *at the receiver* with a filter. There are too many different circumstances to give specific solutions, but Table 17-1 provides some guidance and the task of installing a filter at a television receiver's input is presented as an example later in this chapter.

Table 17-1:	Interference-Reducing Filters	
Filter to Use	*Effect of Filter*	*Type of Interference*
High-pass	Rejects signals *below* the *cutoff* frequency.	Use on FM and TV receivers experiencing interference from shortwave, AM, amateur HF, and CB transmitters.

Filter to Use	Effect of Filter	Type of Interference
Low-pass	Rejects signals *above* the *cutoff* frequency.	Shortwave and amateurs use this for interference from strong local FM or TV transmitters. Use on FM and TV receivers receiving interference from commercial VHF and UHF transmitters.
Notch or trap	Reject a specific *range* of frequencies.	Often used to reject signals from a single nearby paging, FM, or TV transmitter.
Band-pass	Rejects signals *outside* a specific range of frequencies.	Used when multiple interference sources are present.

The *Interference Handbook* published by the American Radio Relay League (www. arrl.org) is an awesome and inexpensive reference, covering all kinds of interference problems relating to radio signals — and the techniques for curing them.

Direct detection

Electronic equipment not designed to receive radio signals can act like a radio receiver, too. You may not think you have antennas connected to your electronics, but look again — every cable and cord you have plugged in acts like an antenna for any radio wave that goes by! The diodes, transistors, and ICs in the circuits connected to the cable all act as crystal radio sets, converting the radio wave to an interfering signal inside your equipment! This is called *direct detection*. A good example might be garbled audio from public-address system picking up a two-way radio transmission, even though it doesn't have a single radio circuit inside.

The solution in direct detection is to install radio wave absorbing *chokes* on the cables to prevent the signals from getting inside your equipment in the first place. The cables can be wound onto a *core* as shown in the "Installing a Split-Core Ferrite Choke" task later in this chapter. Ferrite cores can also be installed on cables that are too big to conveniently wind onto a core.

Choke: An inductor intended to restrict or choke off RF energy in a circuit.

By winding a cable onto a ferrite core, the effect of the ferrite is multiplied with each additional turn through the center of the core. In the task later in this chapter, you'll see how a power cord is wound onto a split core that's available at most electronic component stores.

Avoid causing interference

Even non-radio equipment can cause interference! This is more accurately referred to as *EMI* (electromagnetic-frequency interference) or *RFI* (radio-frequency interference). Signals can leak out of your equipment from an oscillator or clock signal — this is often a problem with digital circuits that operate with high-speed pulses. In this case, you apply the same filtering techniques — but from the *inside!*

What is this ferrite stuff?

Ferrite material is often used to soak up unwanted radio-frequency (RF) signals to prevent interference. Ferrite is a ceramic material that can absorb RF and keep unwanted signals from getting into or out of electronic equipment. When a wire or cable is completely encircled by ferrite, it's as if an inductor (coil) has been added. This acts to block the flow of RF current. Ferrite also dissipates some of the RF energy as heat. Either way, a small amount of ferrite at the right spot can make a dramatic reduction in interference to or from electronic equipment.

Ferrite is available in a number of forms and with different RF-eating properties. The ferrite sold at electronics stores is formulated to be most effective for signals ranging between 1 and 100 MHz. The most common shapes are:

✔ **Toroids are continuous rings.** Cables and wires are wound onto the rings by passing them through the center of the toroid.

(one is visible at the upper right of Figure 17-1). These are also called *cores*.

✔ **Beads are like toroids, but cylindrical.** Two large beads are shown inside the toroid in Figure 17-1. They are slipped on to cables and wires and several may be combined.

✔ **Split-cores are large beads that are sawn in half.** (Don't try it yourself — ferrite is too hard and brittle for cutting without special techniques.) A plastic wrapper holds the two halves together and is secured by a snap. This allows a large cable to be inserted in the center of the bead, even if a connector is already installed on the cable. Smaller cables can be coiled such that the turns pass through the center of the core.

Instead of large ferrite chokes on the cables outside the equipment or low-pass filters in the cables themselves, filtering is performed on the signals before the interfering energy can leave the equipment.

The filtering techniques in this section work in *both* directions! By adding filtering to your device's external connections, you reduce the signals generated by your circuit and also reduce its susceptibility to interference from outside. Adding filtering is just about always a good practice.

If your electronics are radiating signals of any sort, your device is an *unintentional radiator* under the FCC's "Part 15" rules that govern unlicensed devices. Part 15 devices are required to accept any interference they experience or to stop operating if they cause interference to an FCC-licensed station. Commercial Part 15 devices, such as cordless phones and wireless network gadgetry, have this spelled out in their operating manuals or on a sticker right on the equipment. These rules allow useful unlicensed devices to coexist with the more capable licensed radio services.

Figure 17-2 shows four ways to add filtering to the input and output connections of your circuits so you can prevent interfering signals from entering or leaving.

At 17-2(A), two *RF chokes* are shown. These are intended for use in power-supply connections. They're wound with heavy wire and can handle several amps of current. These chokes are very effective at suppressing RF signals up to 10 MHz or so.

At 17-2(B), a *Pi filter* is shown. The name "Pi" comes from the arrangement of the two capacitors and an inductor on the schematic — it looks like the Greek letter π. Depending on the frequency range of interest, the capacitors are from 100 to 1000 pF and the inductor from 0.1 to 10 μH. Some manufacturers also sell EMI filters of this type as small components that resemble capacitors.

The circuit at 17-2(C) is called an RC filter because it's made of a resistor and a capacitor. The capacitor is typically on the output side and is 100 to 1000 pF. The resistor is chosen so the cutoff frequency of the filter is high enough so the desired signals are not affected, but low enough to insure that RF signals are attenuated. The cutoff frequency of the filter is $f_C = 0.159/RC$. (L measured in henries and C in farads)

17-2(D) shows the symbol for a ferrite bead. Ferrite beads can be slipped onto individual wires if a large core is too cumbersome. The beads are typically ¼" in diameter or smaller and several may be used on one wire. Beads of this size work best for signals above a few MHz and on wires carrying minimal current.

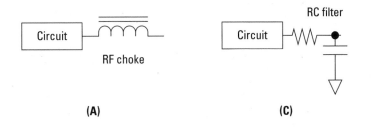

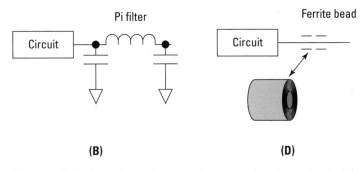

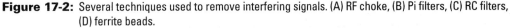

Figure 17-2: Several techniques used to remove interfering signals. (A) RF choke, (B) Pi filters, (C) RC filters, (D) ferrite beads.

Installing a High-Pass Filter

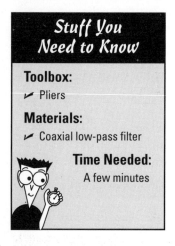

Stuff You Need to Know

Toolbox:
- Pliers

Materials:
- Coaxial low-pass filter

Time Needed:
A few minutes

If your television is receiving interference on many channels from a single strong nearby signal, it's likely that the television receiver is overloading. Luckily, there is a simple fix if the interfering signal is of a frequency *below* that of television signals. The lowest-frequency TV signal is found at channel 2, between 54 and 60 MHz. A high-pass filter reduces *(attenuates)* lower-frequency signals. A typical TV-type high-pass filter is shown in Figure 17-1. (The MFJ Enterprises MFJ-711B is a good choice, www.mfjenterprises.com/products.php?prodid=MFJ-711B). These are easy to install as shown in this task.

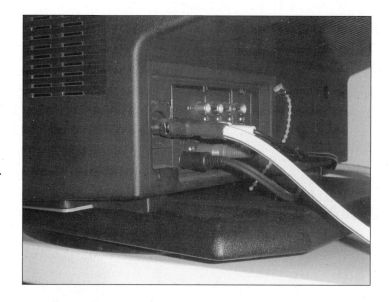

1. Locate the input cable to your TV, DVD player, or VHS recorder as shown in the photo. It will be a Type F connection (see Chapter 8).

WARNING Check with your cable TV or satellite receiver provider before installing any filters in the input cable to the receiving equipment. These systems use different ranges of frequencies than over-the-air TV receivers. Luckily, these systems rarely experience overload. The cables to satellite antennas also carries DC voltages to power the receivers and switches. Installing a filter in the feedline may block or short out the DC power, preventing the system from working.

2. Disconnect the cable by unscrewing the connector.

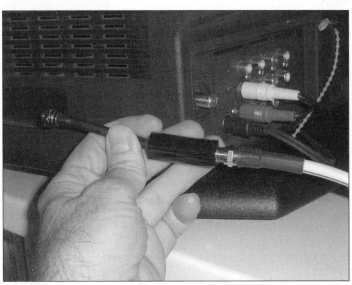

3. Connect the cable to the threaded end of the filter. Seat the connector firmly with the pliers. Do not overtighten — you just want a secure connection.

4. Confirm that the center conductor in the center of the Type F connector at the other end of the filter is not bent or pushed back into the connector. It should extend slightly beyond (or be flush with) the end of the connector's screw-on shell.

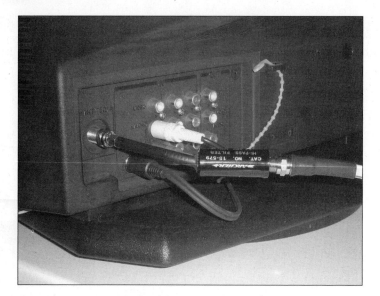

5. Screw the shell of the filter's connector onto the threaded connector on the equipment. Tighten with the pliers.

6. Turn on the equipment to make sure that you're still receiving signals and that their quality is good.

Installing a Split-Core Ferrite Choke

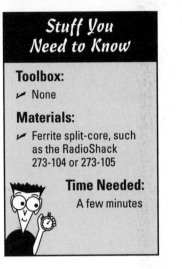

Stuff You Need to Know

Toolbox:
- None

Materials:
- Ferrite split-core, such as the RadioShack 273-104 or 273-105

Time Needed:
A few minutes

It's quite common for RF signals to cause interference in consumer electronics, entering the equipment via the power cord. Plug-in RF filters are available, but can be expensive. A more cost-effective first step is to use a split-core ferrite choke, winding the power cord through it several times.

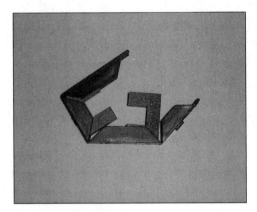

1. Open the split core by releasing the locking catch at one corner of the plastic housing.

2. Starting as close to the equipment as you can, wrap the cord through the center of the core until the center of the core is filled. All turns must be wound in the same direction to avoid canceling the effects of the core.

3. When the core is full, slide the other half of the core onto the cable so the halves meet securely.

4. Keep the halves together by securing the locking catch.

Part VI
The Part of Tens

In this part . . .

The best part of the *Dummies* book, at last — the Part of Tens! With so many ways of building and so many things to be built, one chapter presents a set of ten circuit-building secrets. These little bits of workbench wisdom are intended to help you along the path to success. Then you get ten tips for circuit first-aid — just what you may need when things go wrong away from home.

I've enjoyed writing this book and trying to capture some of the things I've learned in more than 40 years of tinkering and building. As a youngster taking apart TV sets for the parts, I often wished there was a basic guide to the newcomer — and now I've had my chance to provide one for you! As you build up your own set of tips and tricks, don't forget to pass them along to someone else just learning the ropes. Circuitbuilding is more than an activity; it's a community!

Chapter 18

Ten Circuitbuilding Secrets

This chapter contains ten bits of wisdom that you might get from a circuit-building mentor — the kind of guidance that isn't about a specific technique, but generally applies to everything. In fact, these are good guidelines for many hobbies and crafts. They help make and keep circuitbuilding an enjoyable activity that will keep on being enjoyable for years and years.

Be Patient and Alert

Take your time at the workbench — it's not a race! After a long session bent over the soldering iron, take a walk to clear your head and loosen your muscles. When you're troubleshooting and starting to get a little frustrated, that's the time to take a break. Many a project has been ruined or delayed because of working in a hurry or past the point of being alert.

Spring for Quality Tools and Toolbox

Buy the best quality tools you can afford. Some of the author's tools have been with him for almost 40 years! Keep them clean and dry, don't abuse them, and store them in a good tool chest or carrying case. Beware of "grocery-store specials!" A high-quality tool will get the job done better and faster, plus it won't wear out as quickly.

Use Plenty of Light

It's important to be able to see what you're doing! Particularly when working on surface-mount electronics or other miniaturized circuits, good lighting can mean the difference between success and failure. A swing-arm lamp with a high-intensity bulb can put brilliant light exactly where it's needed. A head-mounted LED lamp will also do the job.

Get Good References

Look around the shop of a long-time circuitbuilder and you will probably see a handful of books on a shelf near the workbench — usually a mix of old and new, with the old ones somewhat dog-eared and well-used. Ask why those books are kept handy. You'll need a parts and tool catalog or two (those from the vendors listed in Appendix A are great to have), a couple of design cookbooks (one for analog circuits and one for digital), and handbooks covering your interests. When you find a good one — hang on to it!

Hold On to Your Junk

Here's one of the best-kept secrets — the junk box full of spare, castoff, and reused items. Entire projects can be built from a well-stocked junk box! Start with one for mechanical hardware (screws, nuts, springs, and so on) and another for electronic parts (transistors, resistors, knobs, and so on). Save the conductive foam and bags that your parts are shipped in; you can use them in your junk box, too!

Buddy Up

Get to know other local circuitbuilders. They can be invaluable sources of information and parts. Troubleshooting can also go a lot smoother with a friend around to ask the questions you didn't think of. Have an open house and learn from each other. Attend local flea markets or conventions together. Keep on the lookout for ham radio, computer, and antique radio flea markets — these are usually treasure troves of electronic bargains. The whole will definitely be greater than the sum of the parts!

Test in Steps

As you build a project, stop as you complete sections of circuitry to perform simple checks. Did power get connected to everything it should and nothing it shouldn't? Does the amplifier amplify? Does the filter filter? Does the switch switch? These are much easier to test before the entire circuit or system is built. After everything is put together, any bugs in the finished product can be devilishly hard to isolate, so find as many as you can when things are simple.

Keep a Notebook

Memory being what it is — that is, deficient — workbench and product notebooks can be worth their weight in gold. No, even more! If you wire up a control cable, be sure to write down the color of the wire for each terminal. If you get a new project working, be sure to write down any measurements you made of how it performed. Later on, it will be really nice to be able to go back and compare measurements. Just a spiral-bound notebook of lined paper is fine. Note the date for each entry so you can find them later.

Pass It On

As you learn more and more, take the time to pass that information along to others. If you've become experienced, share what you've learned with demonstrations at a club meeting or other gathering. It doesn't matter if you're not a polished public speaker, just lay out your gear, give a little explanation, and let the questioners lead the conversation. And take the time to help a newbie over the same bumps and obstacles you encountered!

Take Pride in Your Craft

Last, but not least, take pride in the craft of your workmanship! Not all projects need to be works of art, but paying attention to detail and appearance often pays benefits of fewer failures and easier maintenance. Plus, you'll swell with pride when somebody looks over your work and says, "Nice job!"

Chapter 19

Ten Circuit First-Aid Techniques and Supplies

Circuitbuilding skills come in extra handy when you can apply them on the spot to make repairs or rescue a broken gadget. You'll have to be prepared, though, and that's what this chapter is all about. By assembling an electronics first-aid kit, you'll be ready to tackle a lot of the "Murphy bites" that seem to occur at the most inopportune times.

Common Replacement Transistors and ICs

Somewhere in your toolbox, keep a small plastic box with a selection of common transistors and ICs. Many circuits can be made to work, although possibly not at peak performance or under all conditions, with a substitute chip or transistor. Here's a good starter list of handy parts; add more that suit your specialty:

- ✔ Bipolar transistors — 2N3904, 2N3906, TIP41, TIP42;

- ✔ MOSFETs — 2N7000, IRF510;

- ✔ Diodes — 1N4148, 1N4001, 1N4007;

- ✔ Analog ICs — 741, LM324, 7805, 7812, LM386. (Sketch the pin configurations on a piece of paper you keep in the same box.)

Clip Leads

Definitely useful in a pinch, a small bundle of various types of clip leads should be in your tool kit. Have a couple with heavy clips and wire that can handle currents of a few amps. A clip lead with a banana plug on one end can often replace a broken or lost multimeter probe.

Electrical Tape

Every toolbox should have a least one good-sized roll of quality electrical tape, such as Scotch 33+ or 88. Use it to replace heat-shrink tubing in a splice, to insulate a terminal (each layer is rated at 600 V or more), or waterproof a connection. A roll of cheap tape is good to have for holding things together. Electrical tape can also be used to make a temporary repair of small holes in hoses and tubes.

Wire Nuts and Crimp Splices

You will be amazed at what you can achieve just by twisting wires together, even for audio and video signals! Miniature wire nuts — cousins to the ones used for house wiring — work well for signal-level wiring. These don't have the holding power of the larger models, so give them a couple of wraps of tape to keep the connection together. Crimp splices are like butt-splice terminals, but they only have one end. Much like a crimp-on wire nut you can use a crimp splice to join two wires.

Molded Connectors

Call it an example of Murphy's Law: It's practically inevitable that cables with a molded-on connector (one whose body is solid plastic) will fail so close to the molded portion that there isn't room for a real splice. So use a very sharp pair of wire cutters (the pointier the nose of the cutters, the better) to dig into the molded plastic and expose enough wire for a temporary splice. Don't use a knife; it's way too easy to cut into the wire (or yourself) that way. Cutters allow you to work with much more control as you gradually expose the wires.

12V Soldering Iron

A soldering iron that can run off a car's cigarette lighter is occasionally a lifesaver. Keep one in your car in a sturdy bag, along with some solder, electrical tape, a pair of needle-nose and wire-cutting pliers. The power cord for these irons usually has a cigarette lighter plug. Purchase or make an adapter that changes the plug into clips that can be attached directly to the battery. You can also use a soldering iron to punch small holes in thin plastic — it doesn't smell very good, though, so do this in a well-ventilated area.

Clothespin and Rubberband Vises

When you're attempting a field repair, all your cool tools, clamps, and work aids are back at home — which can make soldering or splicing a job for an octopus. The solution is to use wooden clothespins or pliers with rubber bands around the handle as your working vise. These will steady the work piece enough for you to work on it without slipping.

Loose Connectors

Intermittent connections are often caused by connectors with loose contacts. For example, the inner contact of a phono jack can work loose over time, either from many insertion-removal cycles or because the cable is twisted sideways while inserted. To bring this kind of sleeve contact back to life, squeeze its halves together gently with needle-nose pliers until it grips the mating connector pin firmly. If the contact is completely surrounded by a connector's body material, a jeweler's screwdriver can often be slipped between the contact and the body to force the contact towards the center.

Broken Antennas

If the antenna for a VHF or UHF radio (such as a scanner or a Family Radio Service handheld two-way radio) is broken off or lost, substitute a length of wire stuffed into the connector's center contact socket. You'll get best results if the antenna is about ¼-wavelength long at the frequency of interest. Use this formula:

Length (in inches) = 2808 / frequency (in megahertz)

For VHF frequencies near 150 MHz (for example), this type of antenna is about 19 inches long. For FM radios or portable TVs, the broken piece of antenna can be crimped a little with pliers and stuffed back inside the part mounted to the equipment. A miniature hose clamp will hold the two pieces together until the antenna can be replaced.

Dead Rechargeable Batteries

Yet another example of Murphy's law: Rechargeable batteries or power packs can always tell when you don't have a charger handy, can't they? That's usually when they run out of power. In a pinch, of course, you can do a quick-charge on a battery pack from a car's battery — but be careful:

✔ Don't connect the batteries directly to the car battery — they'll overcharge and could become extremely hot or explode!

✔ Use clip leads to connect a brake light or turn-signal bulb in series with the batteries to be charged. The bulb will absorb the initial inrush of current and act as a brake on the charging process.

✔ If you feel the discharged batteries getting warm, stop charging! You can use the partially charged batteries for a while until you can replace them.

Glossary

• A •

Adapter: Connector that allows connectors of different types or families to be connected together.

Algorithm: An equation or procedure that produces a result in a fixed number of steps.

Alternating current (AC): Electrical current that flows in alternating directions.

Ammeter: A test instrument that measures current.

Ampacity: A wire or cable's rated current-carrying ability.

Ampere (A): The unit of measurement of electrical current flow, named for Andre-Marie Ampere, a 19th-century scientist who studied electricity. One ampere = 6.25×10^{18} electrons per second.

Amplification: The process of increasing the voltage, current, or power of a signal.

Amplifier: A circuit that increases the amplitude of a signal.

Amplitude: The size or strength of a signal.

Analog signal: An electrical signal that has continuously varying voltages, frequency, or phase.

Analog switch: A digitally controlled semiconductor switch for analog signals.

Anode: The positive terminal of a diode, thyristor or vacuum tube (also called the *plate* in vacuum tubes).

Astable: A device that has no stable state, oscillating regularly between two or more states.

Attenuate: To reduce the amplitude of a signal.

Audio: Signals with frequencies between a few hertz and 20 kHz.

Autorange: To automatically adjust an instrument's scale so the measured value can be displayed.

• B •

Band-pass filter: A circuit that attenuates signals above and below a specific range of frequencies.

Base: The control terminal of a bipolar transistor.

Battery: A device that converts chemical energy into electrical energy.

Beta: Greek letter β (lower-case) used as symbol for the current gain of a bipolar transistor, the ratio of collector current to base current.

Bias voltage: DC voltage applied across the terminals of a PN junction. Forward bias causes current to flow and reverse bias prevents or reduces current flow.

Bipolar transistor: Three-terminal component with three layers of alternating P- and N-type semiconductor materials (constructed as NPN or PNP).

Bistable: Circuit with two stable states.

Block diagram: A drawing using boxes to represent sections of a complicated device or process and the connections between them.

Blow: Referring to a fuse or wire, to open a circuit by melting from excessive current.

Blow time: The maximum time required for a fuse to open after being subjected to a current higher than the rated current. Fuses can have slow, normal, or fast blow times.

Bridge rectifier: Four diodes configured as a bridge that changes AC to full-wave pulsating DC.

Buffer: Circuit used to provide isolation between an input and output.

Bug: Design error or flaw in a circuit or program.

Bus: For supplying power, a bus is a common connection, often a heavy printed-circuit board trace, wire or strap, for power or ground. For the transfer of data and address information, the bus is a set of connections that carry data or addressing information in a computer.

• C •

Calibrate: Adjust a circuit to perform equivalently to a specified reference or level of performance.

Capacitance: The amount of charge a capacitor stores per volt of charging potential, measured in farads (F).

Capacitor: Electrical component formed by separating two conductive plates with an insulating material (dielectric). A capacitor stores energy in an electric field.

Capacity: A battery's ability to deliver current.

Cathode: The negative terminal of a diode, thyristor, or vacuum tube.

Centi (c): The metric prefix for 10^{-2}, or divide by 100.

Chassis: The metal frame or structure on which electronic equipment is assembled or mounted.

Chassis ground: The common connection for all parts of a circuit that connect to the negative side of the power supply.

Chip: *See* IC.

Choke: Inductor intended to prevent the flow of AC current, usually RF current.

Circuit: A group of interconnected components configured for a specific purpose.

Circuit breaker: A magnetic or bimetallic device that opens a circuit when a fault condition is detected.

Closed circuit: An electrical circuit with an uninterrupted current path.

CMOS (Complementary Metal-Oxide Semiconductor): A combination of an N-channel and a P-channel MOSFET in a single switching circuit.

Coaxial cable: Coax (pronounced *ko*-aks) is a type of feed line with a center conductor inside an outer shield that's often made with flexible braid.

Coil: *See* Inductor.

Collector: One of the output terminals of a bipolar transistor.

Color code: A system in which numerical values are designated by various colors to indicate a component's value.

Component: An individual part or element that performs a designated function within an electronic circuit.

Conductor: A material in which electrons can flow easily.

Core: Material used to concentrate the magnetic field of an inductor.

Current: Flow of electrons in an electrical circuit.

Cycle: One complete repetition of a periodic waveform.

• D •

Data sheet: A complete description of an electronic component's properties and characteristics.

Delay time: The time it takes for a circuit breaker to open after its rated current is exceeded.

Diac: A two-terminal semiconductor used for triggering a triac.

Dielectric: Insulating material in a capacitor or coaxial cable that stores or carries electric or electromagnetic energy.

Digital signal: A signal that has only specific values of current or voltage.

Diode: A device that allows current to flow in one direction only.

Direct current (DC): Electrical current that flows in one direction.

Discrete component: An individual component not combined with others in a common package.

Dissipation: The act of removing or giving off power or energy.

Double-pole, double-throw (DPDT) switch: A switch with two sets (poles) of three contacts, each set containing a common contact that can be connected (thrown) to either of the remaining contacts in that set. Both sets of contacts are mechanically linked to change position at the same time.

Double-pole, single-throw (DPST) switch: A switch with two pairs of contacts (poles) that can be connected together or separated. Both pairs of contacts are mechanically linked to change position at the same time.

Drain: One output terminal of a JFET or MOSFET.

Duty cycle: A measure of the amount of time a circuit is operating over a given interval.

• E •

Earth ground: A circuit connection to the Earth.

Electrolyte: A solution of chemicals that serves to connect the electrodes in a battery or capacitor.

Electrolytic: A type of capacitor whose dielectric is a chemical solution or paste of electrolytes.

Electromotive force (EMF): The electrical force that causes current to flow in a circuit, measured in volts.

Electron: An atomic particle that constitutes the smallest unit of electrical charge.

Electrostatic discharge (ESD): The sudden discharge of static energy, usually from a human body to a piece of electronic equipment. The energy in the discharge can be quite damaging to sensitive electronic components.

Emitter: One of the output terminals of a bipolar transistor.

• F •

Farad (F): The unit of capacitance, named for Michael Faraday, a 19th-century scientist who studied electromagnetism.

FET (Field-Effect Transistor): A three-terminal semiconductor that uses electric fields to control current through its output terminals.

Filter: A circuit that passes signals of some frequencies while attenuating other frequencies.

Frequency (f): The number of cycles per second of an AC waveform, measured in hertz (Hz).

Front-end overload: Interference to a receiver caused by a strong signal that overpowers the receiver's input circuits.

Full scale: The maximum reading an instrument can produce for a specific scale setting.

Full-wave rectifier: A circuit of either two or four diodes that changes AC to pulsating DC at twice the input frequency.

Fuse: A thin metal strip that melts when subjected to excessive current, opening the circuit.

• G •

Gate: The input or control terminal of an SCR, triac, or FET.

Giga: The metric prefix for 10^9, or times 1,000,000,000.

Graticle (or graticule): Scales arranged in a grid overlaid on the screen of an oscilloscope.

Ground: The part of a circuit that acts as a reference for the voltages in that circuit.

Ground plane: Continuous conducting surface at ground potential.

Ground rod: A copper or copper-clad steel rod that's driven into the Earth to provide a ground connection.

• H •

Half-wave rectifier: A circuit consisting of a single diode that changes AC to DC that pulsates at the input frequency.

Heat sink: A metal base or plate onto which one or more components are mounted to absorb, carry away, or radiate the heat generated by the component(s). Overheating may result in the malfunction or destruction of the part(s) generating the heat or might cause damage to other parts of the circuit.

Henry (H): The unit of inductance, named for Joseph Henry, a 19th-century scientist who studied magnetic fields.

Hertz (Hz): The unit representing one cycle per second.

High-pass filter: A filter designed to pass high-frequency signals, while attenuating low-frequency signals.

• I •

IC: Integrated circuit or "chip". A group of components (primarily resistors and transistors) fabricated from a common piece of semiconductor material.

Impedance: The opposition to AC and DC current.

Inductance: A measure of the ability of a coil to store energy in a magnetic field, measured in henries (H).

Inductor (L): A component usually composed of a coil of wire and which stores energy in a magnetic field.

Insulator: A material in which electrons flow with difficulty, preventing current flow.

Interlock: Mechanical safety device that removes power from circuits when an enclosure is opened.

• J •

Jack: Female connector, usually mounted on electronic equipment.

JFET (Junction Field-Effect Transistor): An FET constructed with its gate terminal forming a PN junction with the channel between the output terminals.

Joule (J): The unit of energy.

Jumper: A short length of wire or cable used to connect a pair of terminals or printed-circuit board traces.

Junction, PN: In a semiconductor, the point at which N-type and P-type material come together.

• K •

Kilo (k): The metric prefix for 10^3, or times 1000.

• L •

Layout: Arrangement of components in a circuit.

Lead: Wires by which attachments are made to an electronic component or device.

LED: Light-emitting diode.

Level: *See* Amplitude.

Line voltage: The AC voltage supplied by an electric utility.

Linear amplifier: A circuit whose output signal is an amplified reproduction of its input signal.

Load: A device, component, appliance, system, or machine to which electrical power is applied.

Low-pass filter: A filter designed to pass low-frequency signals, while attenuating high-frequency signals.

• M •

Mega (M): The metric prefix for 10^6, or times 1,000,000.

Micro (μ): The metric prefix for 10^{-6}, or divide by 1,000,000, represented by the Greek letter micron.

Microphone: A device that converts sound waves into electrical energy.

Milli (m): The metric prefix for 10^{-3}, or divide by 1000.

Monostable: A circuit device that has a single stable state.

MOSFET (Metal-Oxide-Semiconductor FET): An FET whose gate is insulated from the channel between the output terminals by an insulating layer of oxide.

Multimeter: An electronic test instrument used to measure current, voltage, and resistance.

Murphy: Fictitious character blamed for causing failures and bugs. Murphy's Law states, "Anything that can go wrong, will go wrong."

• N •

Nano (n): The metric prefix for 10^{-9}, or divide by 1,000,000,000.

National Electrical Code (NEC): A set of guidelines governing electrical safety.

NPN transistor: A bipolar transistor made with a layer of P-type material between layers of N-type material.

Notch filter: A filter that removes or attenuates signals in a very narrow range.

• O •

Offset: A DC voltage or current added to an existing signal.

Ohm: The unit of measurement of resistance symbolized by the Greek letter omega (Ω), named after Georg Ohm, a 19th-century German physicist who discovered the relationship between voltage and current.

Ohm's Law: Ohm's Law states the relationship between voltage, current, and resistance.

Open circuit: An electrical circuit that does not have a complete path for current to flow.

Oscillator: A circuit that produces a continuous sequence of pulses or regular waveforms.

• P •

Parallel: Interface through which data bits flow as sets, such as bytes or words.

Parallel circuit: A circuit in which the same voltage is applied to all components.

Patch: To route a signal from one piece of equipment to another, usually audio or telephone.

PCB: Printed-circuit board.

Period: The time required for an AC signal to complete one cycle; this value is the reciprocal of the signal's frequency, measured in seconds and designated by the letter T.

Photovoltaic effect: The generation of an electrical current by exposure to light.

Pico (p): The metric prefix for 10^{-12}, or divide by 1,000,000,000,000.

Plug: Male connector, usually mounted on the end of cables.

PNP transistor: A bipolar transistor made with a layer of N-type material between layers of P-type material.

Potential: *See* Voltage.

Potentiometer (pot): *See* variable resistor.

Power: The rate at which work is done and measured in watts (W).

Power supply: A circuit that produces DC power output from an AC input voltage.

Probe: Leads used to make contact between a circuit and electronic testing equipment.

• R •

Radio-frequency (RF) overload: *See* Front-end overload.

Radio-frequency interference (RFI): Disturbance to electronic equipment caused by radio-frequency signals.

Rail: *See* Bus.

Range: *See* Scale.

Rectifier: Semiconductor diode, or group of diodes, that changes AC to pulsating DC.

Regulator: Circuit that maintains a steady value of voltage or current.

Resistance: The opposition to DC current flow, measured in ohms.

Resistor: Component that presents a specific amount of resistance.

• S •

Scale: A particular setting of a measuring device's sensitivity.

Scanner: Receiver designed to monitor many channels by automatically switching between them.

Schematic symbol: A symbol that represents an electronic component.

SCR (Silicon-Controlled Rectifier): A semiconductor device with four layers of material arranged PNPN to act as a current switch.

Sensor: Device that converts a physical or electrical measurement into an electrical signal.

Serial: Interface through which data bits flow individually in a sequence.

Series circuit: A circuit in which the same current flows through all components.

Short circuit: An unintended connection between two circuits or two points in a circuit.

Shortwave: Radio-frequency signals between 500 kHz and 30 MHz.

Sine wave: An AC waveform described mathematically by the sine function.

Single-pole, double-throw (SPDT) switch: A switch with one set of three contacts (pole) in which one common contact is connected (thrown) to either of the remaining contacts.

Single-pole, single-throw (SPST) switch: A switch with a single pair of contacts (pole) that are connected together (thrown) or disconnected.

Solder bridge: Unintentional connection made by solder across a gap between two terminals or PCB pads.

Solder mask: Coating applied to PCB surfaces that repels solder to prevent solder bridges.

Square wave: An AC waveform that consists of two alternating voltages or currents.

Surface-mount: Either a type of PCB on which components are mounted by soldering directly onto the conducting surface, or an electronic component that's attached to a PCB's surface without holes.

Switch: Device used to open or close a circuit.

• T •

Temperature coefficient (TC or tempco): The change in the characteristic of a component with temperature and specified in change per degree.

THD: Total Harmonic Distortion, a measure of all signal energy due to distortion of an input signal.

Thermistor: A resistor whose resistance changes in a specified way with the temperature; a thermistor is used as a temperature sensor.

Through-hole: Type of PCB on which components are mounted by inserting their leads into holes that go all the way through the board.

Thyristor: A transistor that remains on after it's activated.

Tie point: Common terminal used for connecting several components or leads.

Transient: Short disturbance in voltage or current.

Triac (triode AC switch): A three-terminal device that functions as two SCRs connected in an inverse, parallel configuration to switch AC current.

Trickle charge: A small current continuously applied to keep a battery at a state of full charge.

Trip: Open a circuit after detecting a fault, such as excessive current.

Twisted-pair: Type of cable in which pairs of wires are twisted together and carry a single signal.

• V •

Varactor: Diode that acts as a variable capacitor whose value changes with the voltage across it.

Variable capacitor: A capacitor whose value may be changed within a certain range.

Variable resistor: A resistor whose value may be changed within a certain range.

Varistor: Protective component whose resistance drops when voltage across it exceeds a specified threshold.

Video: Signals that carry picture or image information.

Volt (V): The unit of electromotive force, named for the 19th century Italian scientist, Alessandro Volta who studied electricity.

Voltage: *See* Electromotive force.

Voltmeter: A test instrument used to measure voltage.

• W •

Watt (W): The unit of power in the metric system.

Waveform: The shape of an AC signal over time; this term usually refers to the shape of a single cycle.

• Z •

Zener diode: A diode that maintains a constant voltage across its terminals after reverse bias reaches a specified voltage.

Appendix A

Circuitbuilding Resources

Like any hobby or craft, circuitbuilding involves a wealth of parts. This appendix provides information on electronic and mechanical components, and where to buy them. In addition, a list of online and print resources is provided so that you can find out more about electronics as you become a more skilled circuitbuilder.

Parts and Pieces: Electronic Components

As a beginning circuitbuilder or electronic-er, you may not be familiar with all the different types of electronic components. You might want to start by reviewing a reference text for electronics such as the ones in the "References" section later in this appendix. The Wikipedia also provides an introduction to electronic components at http://en.wikipedia.org/wiki/Electronic_component.

If you look at a catalog from one of the parts vendors in the Vendors section of this Web page, you may be quite surprised at the number of variations of components, particularly for resistors and capacitors. Why so many types? Which ones are the right choices for a project? To help explain some of the differences, I've provided two more bonus chapters on this book's Web site (www.dummies.com/go/circuitbuildingDIYFD)that discuss the Ps and Qs of Rs and Cs (that is, resistors and capacitors, respectively). Here are some additional Web resources to help you learn about some other types of *passive* (meaning unpowered) components:

- ✔ **Inductors:** Try the application notes and other Web pages available on Coilcraft's Web site (www.coilcraft.com) under "Design Tools."

- ✔ **Switches and connectors:** Catalogs and application notes from the major vendors are often excellent sources of information. Try Switchcraft (www.switchcraft.com) and Amphenol (www.amphenol.com).

- ✔ **Fuses:** The Littlefuse Electronic Designer's Guide (www.littelfuse.com/data/en/Product_Catalogs/EC101-J_V052505.pdf) is large (25 MB) but an excellent reference to save on your hard drive. The first section ("Fuseology") will tell you everything you want to know about fuses.

Semiconductors

Encyclopedias have been written on selecting semiconductor components; everything from simple diodes to complex ICs. Getting into the details of selecting a particular diode, transistor, thyristor (a special type of transistor used as a switch), or integrated circuit is seriously beyond the scope of this book. However, the "References" section later in this appendix has some good places to start reading. Until you get your feet wet with designing your own circuits (metaphorically speaking, of course), it's best to stick with the parts that are specified or provided for your circuits and make direct replacements whenever repair is needed.

Data sheets

The easiest way to find a data sheet on your component is to enter a part number and the phrase "data sheet" into an Internet search engine. When you've found your data sheet, save it on your hard drive (if possible) in a folder called "Data Sheets."

Packaging

Transistors and ICs are often available in several different packaging styles. You may not recognize all of the terms involved, so here is a link (www.smta.org/files/ acronym_glossary.pdf) to an electronic manufacturing glossary that contains definitions of all the common package abbreviations.

It's not uncommon to pick just the right part from a catalog and then find out that it's only available in a package style that won't fit your circuit or tools! (Arrgh!) Luckily, there are adapters for many such situations, such as needed to use a surface-mount part on a through-hole board. These are usually called IC adapters, such as those available at Jameco (www.jameco.com). Enter "IC Adapter" (including the quotes) into the search engine's keyword-search window.

Workshop tips for components

As you start accumulating parts, you'll need to keep them safe and ready for circuit-building action. At the very beginning, you might be able to store your stash in a shoebox, but if you're like most builders, that won't last long; the parts tend to over-flow the space available! Soon you'll have at least one or two parts cabinets with small plastic drawers, all filling with components. So here are a couple of suggestions on how to best manage your growing inventory:

- ✔ Sort resistors by the color of their multiplier band: for example, label the drawer "Brown" for resistors between 100 and 999 ohms.

- ✔ Sort capacitors by type (ceramic, film, tantalum, electrolytic, and so on) and size (1–999 pF, 1–999 nF, 1–99 μF, 100–9999 μF, and so one).

- ✔ Use small, zippered plastic storage bags for transistors.

✔ Keep loose ICs and other static-sensitive components on sheets of styrofoam wrapped in aluminum foil — stick the legs of the ICs through the foil and into the foam.

✔ Office- and craft-supply stores often sell inexpensive folding boxes that are the right size for parts (and are much cheaper than small sets of plastic drawers).

Purchasing tips

Depending on your interests, you'll gradually accumulate a selection of parts that you use frequently. Audio experimenters usually have one set and robot-builders another. In general, all circuitbuilders can benefit from the following tips:

✔ **Read construction articles for parts lists:** Note which are the most popular items in articles describing projects you like.

✔ **Buy in quantity when possible:** Often there's a steep discount for buying (say) ten of something.

✔ **Watch out for minimum-purchase and shipping charges:** Combine your order with those of others; a friend or several club members can help you save big on these charges with combined orders.

✔ **Assortments are a good way to start a stock of parts:** You can save a lot of money over buying in small quantities — and may even get a parts cabinet in the bargain.

✔ **When ordering from a large catalog, look for low-cost equivalent parts:** Often parts with equivalent (or better) specifications are more popular — and cheaper.

✔ **Parts in grab bags are often unusable or unpopular:** Why do you think these parts are in the grab bag?

✔ **Set up customer accounts with your favorite vendor:** Get the vendor's e-mailed newsletters and notifications of specials.

Junk boxes

Unless a component is truly toasted or otherwise unusable, don't throw it away! Toss it into a coffee can or shoebox; a well-stocked junk box is a thing of beauty to the circuitbuilder! Start one for electronic components, one for mechanical hardware, and one for those odd bits and pieces left over from gadget assembly and configuration. At the least, they might save you a trip to the store someday! Salvage, salvage, salvage!

Go Nuts: Mechanical Hardware

Given the tremendous range of mechanical parts from watchmaking to bulldozer repair, what *is* the right stuff for electronics? To get an idea of which parts are useful, look in the Hardware section of catalogs that cater to the circuitbuilder. As with electronic components, assortments are often available — but the ones sold at the local hardware store often contain sizes too large for most electronics work. Here are the sizes of materials most often used in circuitbuilding:

- Pan-head machine screws and nuts: 4-40, 6-32, 8-32, 10-32 and 10-24
- Pan-head sheet metal screws: 4, 6, 8
- Flat washers, split-O lock washers, internal-tooth lock washers: 4, 6, 8, and 10
- Kep nuts (nuts with a plastic insert to grip the screw): 6, 8, 10
- Metric machine screws and nuts: 4, 5, 6, 8mm

For information about mechanical fasteners in general, Bolt Depot has an excellent reference page at www.boltdepot.com/fastener-information.

The Keystone Electronics Company (www.keyelco.com) and Abbatron-HH Smitch (www.abbatron.com) manufacture an amazing array of fasteners and hardware for the electronics industry. Their large selection of useful parts is sold through most electronic distributors, including the following:

- **Solder lugs:** Part lock washer, part solder terminal, these are used for making chassis connections.
- **Spacers and standoffs:** Used to hold assemblies and printed circuit boards away from enclosure surfaces.
- **Grommets and strain relief:** Always use one when a wire or cable goes through a metal wall or panel.
- **Cable clamps:** These hold cables and wire bundles securely in and on equipment.
- **Feet:** You'll need stick-on and screw-on rubber feet for your equipment.
- **Cable ties:** The large ones are found at the hardware store, but small ones (6" and smaller) will have to be purchased at an electronics company.

For the mechanical side of "electro-mechanical," take a look at the Small Parts Company Web site, www.smallparts.com. They have a wonderful selection with no minimum order requirement.

References

This book gives you some good information on how to build circuits, but there's so much more to electronics and circuitbuilding! As you get started, one of the best pieces of advice is for you to start collecting information. This section presents electronics references that can help you delve a lot farther into electronics circuits. Many of these can be found in your local library so you can "try before you buy." These short lists by no means exhaust the possibilities of what's out there, but they can launch you into the world of electronics to explore on your own!

Books

Active Filter Cookbook by Don Lancaster: A source of cookbook-type circuit designs along with full design equations for the experimenter.

The ARRL Handbook by the American Radio Relay League: Covers everything from components to full-blown radio systems and a lot of related electronics (check out the ARRL's extensive publication list at www.arrl.org/catalog).

CMOS Cookbook by Don Lancaster: An excellent "go-to" book for digital logic beginners.

Electronic Circuits 1.1 and *Electronic Circuits 1.2* by Intellin Organization: Collections of useful circuits from a wide variety of electronics.

Electronic Formulas, Circuits, and Symbols by Forest Mims: An excellent workspace reference to keep handy (most titles by Mims are excellent practical texts).

Electronics For Dummies by Gordon McComb and Earl Boysen (Wiley Publishing, Inc.): Written for the hobbyist just getting started in electronics.

Electronics Projects for Dummies by Earl Boysen and Nancy Muir (Wiley Publishing, Inc.): Ten simple projects that will go well with the techniques in this book.

Op-Amp Cookbook by Walter Jung: This is the must-have reference for beginning op-amp users.

Robot Building for Beginners by David Cook: An introduction to applied circuitbuilding in the field of robotics.

Tab Electronics Guide to Understanding Electricity and Electronics by Randy Slone: Teaches electronic principles by guiding the reader through design projects, beginning with an adjustable power supply.

Timer, Op-Amp, and Optoelectronic Circuits by Forrest Mims: An excellent book for the timer circuits alone.

Understanding Basic Electronics by Larry Wolfgang: Bite-sized lessons in learning electronics.

Magazines

Circuit Cellar (www.circuitcellar.com): Many excellent vendors for the more advanced builder.

Make (www.make.com): Many electronic projects and interesting articles.

Nuts and Volts (www.nutsandvolts.com): Articles on anything electronic.

Robot (www.botmag.com): All things robotic.

Web sites

ARRL Technical Information Service (www.arrl.org/tis): Listing of articles and online discussions of electronics used in communications equipment.

Battery University (www.batteryuniversity.com/partone.htm): Lots of information about batteries — their terminology, characteristics, charging requirements, and much more.

Discover Circuits (www.discovercircuits.com/index.htm): Thousands and thousands of circuits and references.

University of Washington Electrical Engineering Store (www.ee.washington.edu/stores): Used by students and faculty of the UW EE Department.

Wikipedia (en.wikipedia.org): While the Wikipedia does take occasional hits in the non-technical areas, its electronics sections are quite reliable.

Tutorials

Colin McCord's Oscilloscope Tutorial (www.mccord.plus.com/Radio/oscilloscope.htm): An excellent introduction to this most useful instrument.

Op-Amp Electronics (www.opamp-electronics.com): Excellent series starting at basic circuits and working up.

U.S. Navy Electricity and Electronics Training Series (www.phy.davidson.edu/instrumentation/NEETS.htm): If you're a U.S. taxpayer, you paid for it, so make use of this excellent and thoroughly tested material to teach yourself!

Web EE (www.web-ee.com/primers/Tutorials.htm#Power%20Conversion): A mix of introductory and advanced electronics tutorials and articles.

Williamson Labs (www.williamson-labs.com): Simple introductory-level tutorials.

Vendor sites

All Electronics (www.allelectronics.com): Surplus and discount electronics.

ARRL Technical Information Service Vendor Cross-Reference (www.arrl.org/tis/tisfind.html): Search for vendors that supply specific types of components or equipment.

Digi-Key Electronics (www.digikey.com): Electronic component distributor.

Jameco (www.jameco.com): Electronics and computer supplies.

Marlin P Jones & Associates (www.mpja.com)**:** Surplus and discount electronics.

MCM Electronics (www.mcmelectronics.com)**:** Source for repair and retrofit parts.

Mouser Electronics (www.mouser.com)**:** Electronic component distributor.

Ocean State Electronics (www.oselectronics.com): Electronic components, tools, test instruments, and kits.

RadioShack (www.radioshack.com)**:** Basic selection of parts, lots of adapters and connectors, and in a mall near you.

Ramsey Kits (www.ramseyelectronics.com): Basic to advanced kits.

Tower Electronics (www.pl-259.com): Connectors and adaptors.

Velleman Kits (www.vellemanusa.com/engine.php): Basic to advanced kits, component assortments.

Index

• *S* •

Want more For Dummies DO-IT-YOURSELF guides?
Check these out

Full-color

DO-IT-YOURSELF
Painting FOR DUMMIES

Discover how to:
- Tackle a wide variety of household painting projects
- Get the right tools for the job
- Create faux finishes that make your rooms look fantastic!

Katharine Kaye McMillan, PhD
Design consultant and coauthor of Home Decorating For Dummies

Patricia Hart McMillan
Interior designer and coauthor of Home Decorating For Dummies

978-0-470-17533-0 • $16.99 • December 2007

DO-IT-YOURSELF
Web Sites FOR DUMMIES

Discover how to:
- Plan your site, register a domain name, and create Web graphics
- Design pages easily with Dreamweaver® templates for portfolio, business, or family sites
- Add a photo gallery, blog, podcasts, videos, or money-making ads

Janine Warner
Author of Dreamweaver® CS3 For Dummies

978-0-470-16903-2 • $24.99 • January 2008

Full-color

DO-IT-YOURSELF
Plumbing FOR DUMMIES

Discover how to:
- Tackle a wide variety of household plumbing projects
- Get the right tools for the job
- Save time and money by doing fixes and upgrades yourself!

Donald R. Prestly
Senior Editor, Handy Magazine

978-0-470-17344-2 • $16.99 • December 2007

DO-IT-YOURSELF
Web Stores FOR DUMMIES

Discover how to:
- Set up a Web store from start to finish
- Plan your store, handle sales, and gain clients
- Do it yourself, with step-by-step instructions and illustrations

Joel Elad
Coauthor of Starting an Online Business All-in-One Desk Reference For Dummies

978-0-470-17443-2 • $24.99 • January 2008

DO-IT-YOURSELF
Circuitbuilding FOR DUMMIES

Discover how to:
- Build electronic circuits from start to finish
- Prepare a project from schematics, then solder and test it
- Work from detailed do-it-yourself instructions with step-by-step illustrations

H. Ward Silver
Author of Ham Radio For Dummies

978-0-470-17342-8 • $24.99 • March 2008

For Dummies DO-IT-YOURSELF guides gives you a new way to get the job done on your own. Packed with step-by-step photos, rich illustrations, and screen shots, these value-priced guides provide do-it-yourselfers like you with the how-to savvy you need to tackle and complete common household or technology projects—whether you're replacing a faucet or building a Web site.

FOR DUMMIES

A Branded Imprint of WILEY

Buy any DO-IT-YOURSELF For Dummies book and get a FREE screwdriver!

If you're a Dummies fan, you'll love our new **DO-IT-YOURSELF** series—heavily illustrated technology and full-color helpful guides that walk you step by step through every task. To help you tackle your projects, we'll send you a **FREE** screwdriver with an interchangeable slotted/Phillips head when you purchase any **Dummies DO-IT-YOURSELF** book.

Instructions for submission:

Submit your name and address on a 3 x 5 card along with the original store-identified cash register receipt dated between 11/1/07 and 9/1/08 showing the purchase of a DO-IT-YOURSELF For Dummies book.

Mail to:

DO-IT-YOURSELF For Dummies
P.O. Box 5960
Clinton, IA 52736-5960

Visit www.dummies.com/go/doityourself for more information!

Get the job done with Dummies DO-IT-YOURSELF!

Offer expires 9/1/08. Requests must be received by 10/1/08. Good only in the USA. Allow 4–6 weeks for delivery. Limit one screwdriver/offer per household.

Wiley, the Wiley logo, For Dummies, the Dummies Man logo and related trade dress are trademarks or registered trademarks of John Wiley & Sons, Inc. and/or its affiliates.

A Branded Imprint of WILEY